MULTI-PARADIGM MODELLING APPROACHES FOR CYBER-PHYSICAL SYSTEMS

Edited by

BEDIR TEKINERDOGAN

DOMINIQUE BLOUIN

HANS VANGHELUWE

MIGUEL GOULÃO

PAULO CARREIRA

VASCO AMARAL

ACADEMIC PRESS
An imprint of Elsevier

Part 2.   Methods and tools

11. Development of industry oriented cross-domain study programs in cyber-physical systems for Belarusian and Ukrainian universities   271

Anatolijs Zabasta, Nadezda Kunicina, Oksana Nikiforova, Joan Peuteman, Alexander K. Fedotov, Alexander S. Fedotov and Andrii Hnatov

List of contributors

Rima Al-Ali
Charles University, Prague, Czech Republic

Vasco Amaral
University of Lisbon, Lisbon, Portugal

Moussa Amrani
University of Namur, Namur, Belgium

Soumyadip Bandyopadhyay
BITS Goa, Sancoale, India

Ankica Barišić
NOVA LINCS Research Center, Lisbon, Portugal

Dominique Blouin
Telecom Paris, Institut Polytechnique de Paris, Paris, France

Didier Buchs
University of Geneva, Carouge, Switzerland

Paulo Carreira
University of Lisbon, Lisbon, Portugal

Moharram Challenger
University of Antwerp and Flanders Make, Antwerp, Belgium

Joachim Denil
University of Antwerp and Flanders Make, Antwerp, Belgium

Nisrine El Marzouki
LIMS Laboratory, USMBA, Fez, Morocco

Ferhat Erata
Yale University, New Haven, CT, United States

Raheleh Eslampanah
University of Antwerp and Flanders Make, Antwerp, Belgium

Alexander K. Fedotov
Energy Physics Department, Physics Faculty, Belarusian State University, Minsk, Belarus

Alexander S. Fedotov
Energy Physics Department, Physics Faculty, Belarusian State University, Minsk, Belarus

Holger Giese

Hasso-Plattner-Institut, Potsdam, Germany

Miguel Goulão

University of Lisbon, Lisbon, Portugal

Andrii Hnatov

Kharkiv National Automobile and Highway University, Kharkiv, Ukraine

Mauro Iacono

University of Campania "Luigi Vanvitelli", Caserta, Italy

Burak Karaduman

International Computer Institute, Ege University, Izmir, Turkey

Geylani Kardas

International Computer Institute, Ege University, Izmir, Turkey

Stefan Klikovits

University of Geneva, Carouge, Switzerland

Nadezda Kunicina

Institute of Electrical Engineering and Electronics, EEF Riga Technical University, Riga, Latvia

Peter Gorm Larsen

Aarhus University, Aarhus, Denmark

Rakshit Mittal

BITS Goa, Sancoale, India

Hana Mkaouar

Telecom Paris, Institut Polytechnique de Paris, Paris, France

Eva Navarro

University of Wolverhampton, Wolverhampton, United Kingdom

Mihai Neghină

Lucian Blaga University, Sibiu, Romania

Oksana Nikiforova

Riga Technical University, Riga, Latvia

Joan Peuteman

KU Leuven, Brugge, Belgium

Ken Pierce

School of Computing, Newcastle University, Newcastle upon Tyne, United Kingdom

Andrejs Romanovs

Riga Technical University, Riga, Latvia

Bedir Tekinerdogan

Wageningen University & Research, Wageningen, The Netherlands

Baris Tekin Tezel

International Computer Institute, Ege University, Izmir, Turkey

Dokuz Eylul University, Izmir, Turkey

Hans Vangheluwe

University of Antwerp and Flanders Make, Antwerp, Belgium

Ken Vanherpen

University of Antwerp – Flanders Make, Antwerp, Belgium

Anatolijs Zabasta

Institute of Electrical Engineering and Electronics, EEF Riga Technical University, Riga, Latvia

Constantin Bălă Zamfirescu

Lucian Blaga University, Sibiu, Romania

CHAPTER 1

Introduction

Bedir Tekinerdogan[a], **Dominique Blouin**[b], **Hans Vangheluwe**[c], **Miguel Goulão**[d], **Paulo Carreira**[d] and **Vasco Amaral**[d]

[a]Wageningen University & Research, Wageningen, The Netherlands
[b]Telecom Paris, Institut Polytechnique de Paris, Paris, France
[c]University of Antwerp and Flanders Make, Antwerp, Belgium
[d]University of Lisbon, Lisbon, Portugal

1.1. Objectives

Truly complex, multidisciplinary, engineered systems, known as Cyber-Physical Systems (CPSs), are emerging in today's reality. Those integrate physical, software, and network aspects in a sometimes adverse physical environment. We can find examples of CPSs in autonomous cars, industrial control systems, robotics systems, medical monitoring, and automatic pilot avionics, to name a few. The cover of the book, for instance, shows another example of a complex CPS, the Maltese Falcon sailing yacht, one of the world's most complex and largest yachts. The act of sailing consists of employing the wind, acting on sails, propelling the craft on the surface of the water in the most different environment scenarios (the Sea conditions can be somewhat unpredictable). Operation of these crafts, with speed and fuel-saving concerns, is traditionally done by skilled skippers and can be appreciated as a sport. Here, like in other sports such as Formula-1, automation stepped into the design, simulation to support the decision and optimise, and operation in both stationary and unsteady conditions. Here, besides the several degrees of freedom of motion, a complex dynamic of forces must be controlled by balancing hydrodynamics, aerodynamics, buoyant and gravitational forces.

As Maltese Falcon, to build such boats is a complicated engineering endeavour with challenging technological and physical constraints (with impact in the used materials, communications, etc.). Maltese Falcon has 88 meters long (289 feet) and can be operated by a single person. To do that, she counts with a wide plethora of actuating devices, including freestanding rotating masts. Also, to support a complex control logic, there is a complex set of sensors with advanced technology such as fiber optical strain net into the spars to analyse real-time loads under sail. The yacht's sophisticated computer software can automatically detect parameters such as wind speed and display critical data to the operator. This autonomy property is characteristic of many CPSs.

Despite the increased adoption and impact of CPSs (where Maltese Falcon is an example), no unifying theory nor systematic design methods, techniques, and tools exist for such systems. Individual (mechanical, electrical, network, or software) engineering disciplines only offer partial solutions. Multi-Paradigm Modelling (MPM) proposes to model every part and aspect of a system explicitly, at the most appropriate level(s) of abstraction, using the most appropriate modelling formalism(s). Modelling languages engineering, including model transformation, and the study of their semantics, are used to realise MPM. MPM is seen as an effective answer to the challenges of designing CPS.

Modelling and analysis are crucial activities in the development of Cyber-Physical Systems. Moreover, the inherent cross-disciplinary nature of CPS requires distinct modelling techniques related to different disciplines to be employed. At the same time, to enable communication between all specialities, common background knowledge is needed.

Anyone starting in the field of CPS will be faced with the need for literature with solid foundations of modelling CPS and with a comprehensive introduction to the distinct existing techniques with clear reasoning on their advantages and limitations. Indeed, although most of these techniques are already used as a matter of common practice in specific disciplines, the knowledge of their fundamentals and application is typically far away from practitioners of another area. The net result is the tendency for CPS practitioners to use the technique that they are most comfortable with, disregarding the technique that would be the most adequate for the problem and modelling goal.

This book is the result of cooperation made possible by the COST Action IC1404 "Multi-Paradigm Modelling for Cyber-Physical Systems" (MPM4CPS), which allowed researchers, institutions and companies from 32 countries to collaborate over four years.

The goal of this book is to serve as a showcase of the research outcomes of the different workgroups of the MPM4CPS network. As such, the book is expected to cover the results on the foundations, formalisms, tools, and educational resources produced within the MPM4CPS network (e.g., case studies made available by the MPM4CPS network). The text will focus on state-of-the-art research and practice knowledge.

The book includes both chapters that discuss experiences from the industry and papers that are more research-oriented. Practitioners will benefit from the book by identifying the critical problems, the solution approaches, and the tools that have been developed or are necessary for model management and analytics. Researchers will benefit from the book by identifying the underlying theory and background, the current research topics, the related challenges, and the research directions for the model management and analytics. The book will also help graduate students, researchers, and practitioners to get acquainted with recent research outcomes of the MPM4CPS network.

1.2. Outline of the book

The book consists of three basic parts: an ontological framework for MPM4CPS; methods and tools; and, finally, case studies. Part 1 provides the ontological framework for

MPM4CPS and includes Chapter 2 to Chapter 5. The ontology framework is decomposed into four sub-parts: a shared ontology to capture concepts that are needed by the other ontologies but that do not pertain to their domains, an ontology for CPS, an ontology for MPM and an integrated ontology for MPM4CPS. For deriving the ontologies, a thorough domain analysis process has been carried out that has focused on key selected primary studies on MPM4CPS. Part 2 describes the methods and tools that are used to develop and analyse CPSs. Chapter 6 to Chapter 9 cover these topics. Part 3 of the book considers case studies which are addressed in Chapter 10 and Chapter 11.

1.2.1 Part 1 – Ontological framework

Chapter 2 introduces the modelling approach and tools that have been chosen to define the ontology, to both motivate and clarify the methods and the actual directions that led the research effort. It also provides a description of the shared ontology whose concepts are used to frame in a more general context some of the more specific notions of the other ontologies. Then, the chapter presents a brief description of the examples we adopted as references to start exploring the field, to glance at our vision of the domain and to guide the reader through our exploration path.

The ontology for CPS is described in Chapter 3. Hereby, a feature modelling approach is adopted that explicitly models the common and variant features of a CPS. Each feature of the resulting feature model is described in detail. The resulting feature model shows the configuration space for developing CPSs. The two case studies on CPS that were introduced in Chapter 2 are used to derive the concrete CPS configurations.

Chapter 4 presents the ontology for MPM specified using the Web Ontology Language (OWL). The chapter first presents a thorough state-of-the-art treatment of MPM's core notions, multi-formalism and model management approaches, languages, and tools, which is an essential component to support MPM. Model management approaches are characterised according to their modularity and incremental execution properties as required to scale for the large complex systems we face today. Subsequently, an overview of the MPM ontology is developed, including the main classes and properties of the ontology. Usage of the MPM ontology is illustrated for the two case studies introduced in Chapter 2.

Chapter 5 integrates the results of the previous chapters by presenting an integrated ontology for MPM4CPS. Hereby, the chapter also elaborates on and integrates the *Shared*, *CPS*, and *MPM* ontologies by providing cross-cutting concepts between these domains. It formalises notions such as model-based development processes, their employed viewpoints supported by megamodel fragments and the CPS parts under development covered by these viewpoints. It finally introduces some ongoing work at the heart of MPM4CPS on the formalisation of modelling paradigm notions in the more general context of engineering paradigms.

1.2.2 Part 2 – Methods and tools

Chapter 6 presents the two-hemisphere model-driven (2HMD) approach for enabling the composition of CPSs. The approach assumes modelling and the use of procedural and conceptual knowledge on an equal and interrelated basis. This differentiates the 2HMD approach from pure procedural, purely conceptual, and object-oriented approaches. This approach may be applied in the context of modelling of a particular business domain as well as in the context of modelling the knowledge about the domain. Cyber-physical systems are heterogeneous systems that require a multi-disciplinary approach for their modelling. Modelling of cyber-physical systems by the 2HMD approach gives an opportunity to compose and analyse system components to be provided transparently and components actually provided, thus identifying and filling the gap between desirable and actual system content.

In Chapter 7, the authors illustrate how a co-simulation technology can be used to gradually increase the details in a collaborative model (co-model) following a discrete event first" (DE-first) methodology. In this approach, initial abstract models are produced using a discrete event (DE) formalism (in this case, VDM) to identify the proper communication interfaces and interaction protocols among different models. These are gradually replaced by more detailed models using appropriate formalisms, for example, continuous-time (CT) models of physical phenomena.

Chapter 8 presents an agent-based CPS development approach using a domain-specific modelling language, SEA ML++. The paper elaborates on intelligent agents, which are software components that can work autonomously and proactively to solve the problems collaboratively. Agents can behave in a cooperative manner and collaborate with other agents constituting systems called Multi-agent Systems (MAS). Intelligent software agents and MASs can be used in the modelling and development of CPSs. In this chapter, the authors discuss how SEA ML++ is used for the design and implementation of agent-based CPSs. An MDE methodology is introduced in which SEA ML++ can be used to design agent-based CPS and implement these systems on various agent execution platforms. As evaluating case study, the development of a multi-agent garbage collection CPS is taken into consideration. The conducted study demonstrates how this CPS can be designed according to the various viewpoints of SEA ML++ and then implemented on JASON agent execution platform.

Chapter 9 focuses on hybrid systems modelling, which is essential for CPS. The expressiveness for hybrid systems modelling allows for the definition of highly complex systems that merge discrete state-based transitions systems with continuous value-evolutions for variables. That way, cyber-physical systems can be modelled in all their intricacies. However, this expressive power comes at a downside of complex models and undecidable verification problems even for small systems. In this chapter, the authors present CREST, a novel modelling language for the definition of hybrid systems. CREST merges features from various formalisms and languages such as hybrid automata, data flow programming, and internal DSL designs to create a simple yet powerful language for modelling resource flows within small-scale CPS such as automated gardening applications and smart homes. The language provides an easy-to-learn graphical interface and is supported by a Python-based tool implementation that allows the efficient modelling, simulation, and verification of CPS models.

1.2.3 Part 3 – Case studies

Chapter 10 describes the design and development of an IoT- and WSN-based CPS using MPM Approach for a Smart Fire Detection Case Study. The system is developed using the Internet of Things (IoT) components and Wireless Sensor Network (WSN) elements. The proposed system is composed of different hardware parts, software elements, computing components, and communication technologies, resulting in a complex system considering both its structure and behaviour. The chapter elaborates on different phases of the development process, including requirement analysis, design, modelling and simulation, and implementation. To present the MPM approach during these phases, the Formalism Transformation Graph and Process Model (FTG+PM) is utilised, and all the involved artifacts and model transformations are described. This helps to provide the data flow and the control flow of the system development in the PM. Further, the analysis of the FTG shows the possible improvements for the system by finding the critical manual transformation to be (semi-)automated.

Chapter 11 presents the development of industry-oriented cross-domain study programs in CPSs for Belarussian and Ukrainian Universities. The paper aligns with the targets of the European COST Action MPM4CPS project, which also considered the dissemination of the results in the educational context. The chapter presents the process and results for creating the base for a European Master and Ph.D. program in MPM4CPS involving European Leading Universities. Further, it elaborates on setting up the respective discipline road-map facing the challenge to develop a mutually recognised cross-domain expertise-based study program in CPS. One of the challenges in the development of study programs is bridging the gap between industry needs and educational output, in terms of training the prospective researchers and engineers in the CPS field. The success of MPM4CPS encouraged EU partners to apply knowledge and methods, developed by the COST Action at an ERASMUS+ project, to validate in practice its viability aiming to develop the industry-focused curricula at the partners' universities of Belarus and Ukraine. The chapter discusses how the COST team efforts towards an analysis of tendencies, industry needs, and acquiring best education practices have been applied by the ERASMUS+ team to create industry-focused cross-domain study programs in CPS for the partners' universities of Belarus and Ukraine.

1.3. Acknowledgements

This book was supported by the COST Action IC1404 Multi-Paradigm Modelling for Cyber-Physical Systems (MPM4CPS), COST is supported by the EU Framework Programme Horizon 2020.

INESC-ID authors were supported by national funds through FCT (Fundação para a Ciência e a Tecnologia) under contract UID/CEC/50021/2019. Also, NOVA authors were supported by NOVA LINCS Research Laboratory y (Ref. UID/CEC/04516/2019) and bilateral project Portugal-Germany, "Modelação de Sistemas Sócio Ciberfísicos", Proc. 441.00 DAAD.

Telecom Paris authors were partially supported by the US Army Research, Development and Engineering Command (RDECOM).

Finally, we would like to thank all the authors and contributors of the chapters.

PART 1

Ontological framework

CHAPTER 2

An ontological foundation for multi-paradigm modelling for cyber-physical systems

Dominique Blouin[a], **Rima Al-Ali**[b], **Mauro Iacono**[c], **Bedir Tekinerdogan**[d] **and Holger Giese**[e]

[a]Telecom Paris, Institut Polytechnique de Paris, Paris, France
[b]Charles University, Prague, Czech Republic
[c]University of Campania "Luigi Vanvitelli", Caserta, Italy
[d]Wageningen University & Research, Wageningen, The Netherlands
[e]Hasso-Plattner-Institut, Potsdam, Germany

2.1. Introduction

The domain of Cyber-Physical Systems (CPSs) arises from the natural technological evolution of control systems, the increased reliability and speed of modern networks and the possibility of obtaining sophisticated coordination, supervision and control of complex, possibly distributed and interacting systems by means of software. CPS conjugate in a single system a computer-based (cyber) part, characterised by discrete time behaviour and isolation from the physical world when non interacting, dominated by software and by algorithmic descriptions, and a physical part, operating in the real, continuous time and in the concrete, tridimensional space, without isolation from the external world and dominated by kinematics, dynamics and all possible undesired influences of natural phenomena.

The complementarity between the two components may allow a prompt management of the physical one by means of a reactive context management and a proper drive of actuators, given an appropriate sensorisation of the physical component: the cyber component may provide unprecedented intelligence, beyond any previous conventional possibility, to the system, assuming that proper design techniques are available to think and develop it as a whole.

Developing an appropriate design technique is very challenging. Variety in components, diversity in specification, criticality in requirements, heterogeneity in the time structure behind the cyber and the physical components and interdependence between different aspects and specifications for the overall system and for its components make the complexity of the design process manageable only if the conceptual approach can encompass every specific aspect of each part of the system while keeping internal coherence in front of modification in the design choices for a component or a subsystem.

Multi-Paradigm Modelling Approaches for Cyber-Physical Systems
https://doi.org/10.1016/B978-0-12-819105-7.00007-6

In fact, as different paradigms coexist in CPS, an appropriate design technique must be aware of the involved paradigms, and actually leverage them to encompass the nature of each component by bending the design process to embrace the most natural modelling paradigm at components level and at each abstraction level in which the overall system can be observed to cope with specific aspects of the general problem.

On these premises, we present in this book parts of the results of a wide cooperative research effort that has proposed Multi Paradigm Modelling (MPM) as a key approach to provide a sound foundation for design processes targeting CPS. MPM supports the coordinated use of different modelling paradigms, model transformation techniques, compositional modelling approaches to shape the target system starting from the speci-fication phase to the validation phase of the design cycle, leveraging different modelling techniques and approaches to evolve the design through models suitable for all the needs that design implies, including the satisfaction of non-functional specifications, such as correctness and performances, and their evaluation from the earliest phases of the design and development process. A MPM approach exploits different tools, different frameworks, different mathematical or formal foundations: consequently, the definition of such an approach suitable for CPS requires a careful exploration of the state of the art of modelling techniques and tools to understand how they are structured, which purposes they serve, what goals they intend to achieve, how they allow one to achieve the goals, and the definition of a common framework for the integration of the seman-tics they use to describe models, results, syntactic aspects, evaluation or generation or transformation processes. Such an exploration led to an ontology-based description of the domains of MPM, CPS and MPM4CPS, in order to map MPM concepts to the CPS domain and to understand how MPM approaches should be shaped to better fit the needs of a CPS design process.

This chapter introduces the modelling approach that has been chosen to define the ontology, in order to both motivate and clarify the methods and the actual directions that led the research effort. Readers will also find a presentation of the overall architecture of the ontological framework, divided into four ontologies including a shared ontology to capture concepts that are needed by the other ontologies but that do not pertain to their domains. This shared ontology is also presented as it supports the other ontologies of the other chapters. Then a brief description of the examples we adopted as references to start exploring the field is presented, in order to provide a glance on our vision of the domain and to guide the reader through our exploration path.

The next chapters will then present in detail our ontology for CPS, to give a precise connotation of what our design target looks like, our ontology for MPM, to present the results of our analysis and provide the reader with a comprehensive knowledge of the state of the art of contemporary modelling culture, and the integration between the two, which is the final result of our work.

The cooperation behind these results has been made possible by the COST Action IC1404 "Multi-Paradigm Modelling for Cyber-Physical Systems" (MPM4CPS), which allowed the collaboration of researchers, institutions and companies from 32 countries over 4 years.

2.2. Ontology development approach

This section presents the approach we have followed for the development of our ontological framework. It required one to analyse the domains of CPS, MPM and their joint use for the development of CPS with MPM as the MPM4CPS integrated domain. The workflow for the ontology development approach is shown in Fig. 2.1.

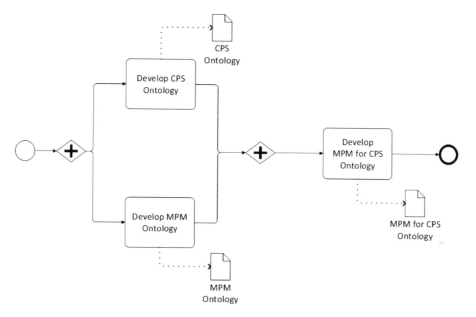

Figure 2.1 Ontology development approach.

Developing these ontologies was performed following an exploratory modelling mode supporting our domain analysis process. Therefore in the following section we briefly introduce this modelling mode. Then we present the followed domain analysis process and finally we introduce the two concrete modelling formalism and tools that were used to capture the domains.

2.2.1 Modelling modes

According to [1], modelling can be categorised into *exploratory* and *constructive* modes each following a different school of thought. Each mode has a different goal with different characteristics employed in order to achieve the goal. Such characteristics are listed in Table 2.1.

Table 2.1 Modelling schools of thought (adapted from [1]).

Property	Explanatory	Constructive
Goal	Understanding	Building
Function	Descriptive	Prescriptive
World View	Open World	Closed World
Growth	Bottom-up (Instances First)	Top-down (Types First)
Typing	Structural	Nominal
Levels	Two-level	Multi-level
Models	E.g. using OWL	E.g. using UML
Logic	Description Logic (DL)	First-order Logic

Exploratory modelling aims at understanding a domain *problem* by providing a description for it, often in the form of a classification. It typically makes use of a bottom-up approach where instances of the domain are studied and typed according to their structure (properties). It therefore assumes an *open world* where not all types are known in advance, but are rather discovered as new instances are classified. It typically uses modelling languages such as the Web Ontology Language[1] (OWL) to specify the classification and Description Logic (DL) to reason about it.

Conversely, *Constructive* modelling aims at building *solutions* for a domain by prescribing nominal types for all elements of the domain. It therefore assumes a *closed world* following a top-down approach where all types of instances are known via an instantiation relation. It uses nominal typing and is supported by modelling languages such as UML and first-order logic via constraint languages such as OCL.

Given the objective of this work, both of these modelling modes are actually needed. It is well known that a sound understanding of a problem to be solved is a prerequisite for building a good solution to the problem. Therefore, a solid understanding of existing CPSs and the way they are developed with models is required to develop a solution to properly relate / combine modelling languages and techniques for CPS development, which was the objective of working group 1 of the MPM4CPS COST action.

In this work, however, we are only concerned with the first phase; exploratory modelling. The objective is that the developed classification of the ontological framework can later serve as a basis to engineer a model management framework using constructive modelling in order to properly Relate / Combine Modelling Languages and Techniques for CPS development. Therefore, this first part of this book on foundations of MPM4CPS is dedicated to presenting the developed MPM4CPS ontologies and the next constructive modelling step is left as future work.

---

[1] https://www.w3.org/OWL/.

2.2.2 Domain analysis process

In order to define the MPM4CPS ontologies, we have followed a domain analysis process inspired from [2–4]. This process consists of identifying, capturing and organising knowledge about the problem domain with the purpose of making it usable when developing new systems for this domain i.e. using constructive modelling. Fig. 2.2 shows an activity diagram for a typical domain analysis process. The main activities of the process consist of *Domain Scoping* and *Domain Modelling*.

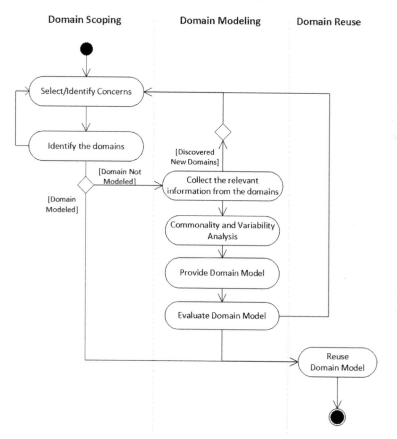

Figure 2.2 Common structure of domain analysis methods (based on [3]).

The purpose of domain scoping is to determine the perimeter of the domain of interest as well as the stakeholders and their goals. Modelling a domain usually requires a lot of effort and therefore, it is important to limit the scope of the domain to only what is required given the needs of the identified stakeholders.

Domain modelling is the activity of building a domain model. Such domain is defined as an area of knowledge characterised by a set of concepts and terminology understood by practitioners of that domain. Building a domain model therefore consists of identifying the relevant concepts of the domain and their relationships for the scope determined during domain scoping. Typically, a domain model is determined by analysing the commonality and variability of the domain concepts. Then a domain model can be specified using dedicated formalisms such as ontological languages, object-oriented languages, algebraic specifications, rules, and conceptual models.

One popular approach for domain modelling consists of using a feature modelling language to capture domain models. Features are properties that are relevant to some stakeholder and used to discriminate between common and specific characteristics of objects. A feature model then defines object features and their dependencies. Feature modelling is widely used to model the commonality and variability of a particular domain or product family.

Besides feature modelling, another technique supporting domain analysis consists of ontology modelling. A commonly accepted definition of an ontology is "*an explicit specification of conceptualisation*" [5]. An ontology represents the semantics of concepts and their relationships using some description language. It can be used to automatically reason about a domain model / classification in order to detect potential inconsistencies given a set of individuals from which the classification was inferred. In this view, feature modelling is also a concept description technique but it focuses on modelling the commonality and variability aspects of a domain. As such, it can be seen as a view on ontologies [4] but without the reasoning capabilities.

For this work, both feature and ontology modelling were used. As explained in Chapter 3, feature modelling was initially used to capture the CPS domain as it is more appropriate for system / product modelling. However, in order to benefit from the stronger analysis capabilities of ontologies, all domains of the framework have in the end been represented as a set of well-integrated ontologies. In order to achieve this integration, a model transformation between feature and ontology models for the languages and tools that were used in the project was developed to automatically convert the CPS feature model into an ontology. This transformation and the choice of language and tools for feature and ontology modelling is further detailed in the next section.

2.3. Modelling languages and tools

The ontology development approach that has been previously introduced was supported by two different languages and their respective tools for feature and ontology modelling. Feature modelling was performed using the FeatureIDE[2] tool and its feature modelling language specified as an XML schema. Ontology modelling was performed using the

---

[2] http://www.featureide.com/.

Web Ontology Language[3] (OWL) expressed as an XML schema and its mainstream tool Protégé.[4]

2.3.1 Feature modelling and FeatureIDE

As mentioned earlier, feature modelling has been used for the CPS ontology to create an initial model as it was thought to be better suited for the CPS domain. Feature models are usually represented as feature diagrams (or tables). A feature diagram is a tree with the root representing a concept (e.g., a cyber-physical system) and its descendant nodes representing the concept features. Relationships between a parent feature and its child features (or sub-features) can be categorised as follows:

- Mandatory—the child feature is required.
- Optional—the child feature is optional.
- Or—at least one of the child features must be selected.
- Alternative (xor)—exactly one of the child features must be selected.

Sets of features can be specified according to these categories to describe product configurations and product lines. Furthermore, constraints can be defined to further restricts the possible selections of features to define system configurations. The most common feature constraints are:

- A requires B—The selection of A in a product implies the selection of B.
- A excludes B—A and B cannot be part of the same product.

The FeatureIDE[5] tool was used to capture the feature model. It has been developed since 2005 and is one of the most mature tools for feature modelling. It is open source and integrated into the Eclipse platform. Since 2014, a commercial supported version is available besides the free version, which includes basic functionalities for feature modelling such as a feature model editor, a feature constraints editor and a feature model configuration editor. A feature configuration model in which the features of a feature model can be selected must be provided for each modelled CPS. Some analyses can be performed by the tool to determine if a configuration is conformed to its feature model.

2.3.2 Web Ontology Language (OWL) and Protégé

OWL is standardised by the World Wide Web Consortium (W3C) and Protégé, which is developed at Stanford University was one of the first tool to implement OWL. Due to their levels of maturity and reasoning capabilities, OWL and Protégé were selected as unifying language and tool to capture the entire set of ontologies of the framework. Protégé has been used for about 20 years for modelling complex domains such as the

---

[3] https://www.w3.org/OWL/.
[4] https://protege.stanford.edu/.
[5] http://www.featureide.com/.

classification of biological species. In addition, there are several existing OWL ontologies that can potentially be reused for this work, such as the SysML QUDV (Quantities, Units, Dimensions and Values)[6] and the QUDT2 (Quantities, Units, Dimensions and Types version 2) for units modelling, or the MAMO (Mathematical Modelling Ontology), which are readily available from repositories such as AberOWL[7] or the Protégé Ontology Library.[8]

The semantics of OWL is mainly based on set theory.[9] Similar to many metamodelling languages such as UML class diagrams, OWL allows the specification of *classes*, *data types* and *properties*. A class specifies a set of individuals while a data type specifies a set of primitive data such as numbers and strings. However, OWL is much more expressive than UML. For example, following set theory, classes can be declared disjoint from each other. In addition, classes can be defined either via their extension or their intension. *Defined* classes express a set of individuals via the set extension or, in other words, via an exhaustive enumeration of the individuals of the set. Classes can also be defined as the combination of other classes via the traditional *union*, *intersection* and *complement* set operators. Finally, a class can also be expressed as a *property restriction*, which consists of an anonymous class representing the set of all individuals that satisfy the given restriction. Such restrictions typically constrain the values of the characteristics of individuals.

Properties are binary relations between individuals. A property has a *domain* that restricts the classes of the individuals from which it origins and a *range* that sets the classes of the targeted individuals. Again, OWL properties are much more expressive than UML properties as they can have many characteristics such as being *functional*, *transitive*, *symmetric*, *asymmetric*, and *reflexive*. Like for classes, an OWL property can be the *subproperty* of another property. Formally, this means that if property 2 is a subproperty of property 1, then the set of individuals or instances paired by property 2 (the property's extension) should be a subset of the extension of property 1.

All these notions contribute to the rich semantics of OWL therefore supporting reasoning on the ontology in order to detect inconsistencies or to infer logical consequences and automatic classification from the ontology individuals. To this end, several plugins that can be integrated in Protégé are readily available for reasoning on OWL ontologies.

Similarly, several plugins for visualising an ontology in the form of different kinds of diagrams, matrices and other forms are also available. For this work, we mainly used the OntoGraf plugin.[10] In the OntoGraf notation, classes are represented as rectangles

---

[6] http://www.omgwiki.org/OMGSysML/doku.php?id=sysml-qudv:qudv_owl.

[7] http://aber-owl.net/.

[8] https://protegewiki.stanford.edu/wiki/Protege_Ontology_Library#OWL_ontologies.

[9] https://www.w3.org/TR/owl-ref/.

[10] https://protegewiki.stanford.edu/wiki/OntoGraf.

with a filled yellow circle on the left side (Fig. 2.4). Subclass relationships are depicted as solid line arrows. As shown in Fig. 4.13, individuals are represented as rectangles with a filled purple diamond on the left side of the rectangle. Individual-to-class relationships are represented as solid line arrows.

As can be seen in Fig. 2.4, in OWL all classes are subclasses (subsets) of a generic top level `Thing` class provided by default in Protege. In addition, it is common practice to separate classes that represent the concepts of the specific domain(s) under study from other utility classes that may be used by that domain but that are not domain concepts. For this reason, as illustrated in Fig. 2.4, under the `Thing` class a `DomainConcept` class is provided to separate these different kinds of elements. This top level domain concept class is further subdivided into the different sub-domains of the ontology, such as `ParadigmDC`, `ProcessDC`, etc.[11]

2.3.3 Integrating feature modelling and OWL

In order to integrate all ontologies into a common technical framework, the FeatureIDE CPS models have been converted into OWL models. This integration is required so that the viewpoints involved at the different stages of development processes that relate the employed formalisms and tools with the parts of the modelled system can be represented in OWL. Furthermore, another advantage of this integration is to benefit from the stronger analyses capabilities of OWL tools.

Both FeatureIDE and OWL models are stored as XML files and their XSD (XML Schema Definition) are readily available. In order to use model transformation tools for converting the FeatureIDE models into OWL, the FeatureIDE and the OWL XSDs were first converted into Ecore metamodels on which most model transformation tools operate. This was achieved almost completely automatically thanks to the Ecore XSD importer; only minor modifications of the generated Ecore classes were required to have correct metamodels for both FeatureIDE and OWL. Such Ecore models are automatically serialised as XML files and are also readable by the FeatureIDE and Protégé tools.

The mapping we employed for mapping feature models to OWL models follows the work of [6]. The detailed specification of this transformation can be found in the deliverable D1.2 of the MPM4CPS COST action project [7].

2.4. Ontology architecture

We present here the overall architecture of the ontological framework serving as foundation for understanding MPM4CPS. In developing this framework, several best practices

---

[11] These sub-domains, which belong to the shared ontology are presented in more detail in Section 2.5.

in ontology modelling have been followed.[12] The first three of them relate to scoping, reuse and modularity.

As already mentioned earlier, developing an ontology for a domain requires significant effort to understand this domain. Therefore, scoping the domain to exactly what is needed given the objectives of the exploratory modelling activity is essential. Furthermore, another way to reduce the modelling effort is to reuse as much as possible existing ontologies for the domains of interest. Finally, in order to ensure that the developed ontologies can be easily reused, developing modular ontologies is essential.

The application of these best practices led to the architecture of Fig. 2.3 for the framework ontologies. In this figure, ontologies are depicted as solid rounded rectangles and labelled arrows indicate different relationships between them.[13] Our framework is divided into three layers. The leftmost layer includes OWL classes and object properties to capture the CPS, MPM and MPM4CPS domains and to be instantiated by the individuals of the other two layers. The *import* arrows indicate that a source ontology makes use of some of the elements contained in the target ontology. In addition, it was found that some classes were needed that did not specifically belong to any of the CPS, MPM and MPM4CPS domains. Therefore, a fourth ontology was developed, namely the *Shared* ontology, to capture these concepts. This ontology can then be used by any of the CPS, MPM or MPM4CPS ontologies. Besides, the CPS ontology may reuse other existing ontologies such as the SysML QUDV (Quantities, Units, Dimensions and Values)[14] ontology, which has already been modelled with Protégé and is therefore readily available. Units systems are indeed an important aspect of the physics part of CPSs, and we can benefit from existing modelling efforts for this domain.

The library layer in the middle of Fig. 2.3 provides individuals of the aforementioned classes such as the catalog of formalisms, modelling languages and tools for CPS modelling, which instantiates the MPM4CPS ontology and its imported ontologies. This catalog ontology was developed during the MPM4CPS COST Action and used to automatically generate the major part of the report on the State-of-the-art on Current Formalisms used in Cyber-Physical Systems Development [9]. In this way, classes and instances can be kept consistent with only minimal coordination efforts. In addition, a SI units system instantiating the QUDV ontology can be provided to support MPM4CPS modelling.

In the third column, examples of CPSs and CPS development environments employing MPM are depicted. The modelling of these CPS development environments was used to refine and validate the MPM4CPS classification following our bottom–up exploratory modelling mode. These examples will be presented in more detail throughout the next chapters of this first part on foundations for MPM4CPS to illustrate the

---

[12] https://www.mkbergman.com/911/a-reference-guide-to-ontology-best-practices/.
[13] The Protégé OWL files can be obtained from [8].
[14] http://www.omgwiki.org/OMGSysML/doku.php?id=sysml-qudv:qudv_owl.

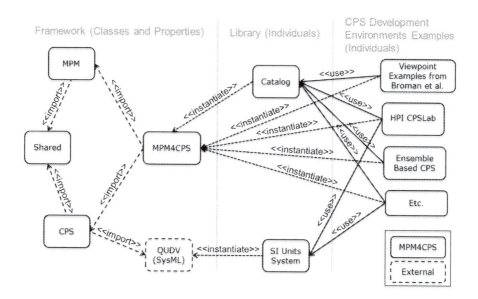

Figure 2.3 Overview of the structure of the MPM4CPS ontology framework.

ontology. As shown in Fig. 2.3, these examples instantiate the MPM4CPS ontology and its imported ontologies and also use some of the languages and tools listed in the aforementioned catalog.

2.5. Shared ontology

We present here the previously introduced *Shared* ontology in greater detail, so that it can be referred from the other chapters presenting the CPS, MPM and MPM4CPS ontologies. The purpose of this ontology is to define notions that do not specifically pertain to the CPS nor the MPM domain but that are required by any of them. In placing these notions in this shared ontology, they can be used by all other ontologies. In addition, the shared ontology also defines generic notions providing a frame for the entire set of ontologies of the framework that can be refined to capture notions for the more specific domains of CPS, MPM and MPM4CPS.

Fig. 2.4 shows an overview of the classes of the shared ontology displayed as a diagram of the OntoGraf plugin (Section 2.3.2). At the root of the ontology, a set of domain concept classes (whose name ends with the DC suffix) are provided to organise classes into the subdomains of *linguistic*, *workflow*, *project management*, *architecture* and *paradigm*. By default, we note that these domains are not made disjoint from each other and therefore their classes are free to belong to more than one subdomain by subclassing the corresponding domain concept classes.

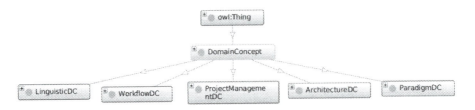

Figure 2.4 Overview of the shared ontology subdomains.

In the following, for each of the subdomains we first briefly introduce some state of the art that served in defining the domain notions, followed by the definition of their classes and properties. It should be noted that it is not the purpose of the subdomains of the shared ontology to provide a complete coverage of the domains. This would have required too much effort. However, defining this initial subdomain classification provides a frame to organise other classes and subdomains of the framework, as will be illustrated in the other chapters presenting notions refining classes and properties of these domains. Besides, it points the way for further development of the ontology by completing the subdomains as needed by future evolution of the framework.

A thorough description of each of the classes and object properties of the ontology would take too much space. Instead, we chose to only describe the main classes and properties, especially those that are used by the description of the examples. In addition, we do not mention all characteristics of classes and properties. The complete specification can be found from [8] where links for accessing the OWL models are given. Besides, throughout the description of the ontologies in this chapter and in Chapter 4 and Chapter 5, we adopt the convention to write exact names of classes and object properties using a `Typewriter-styled` font.

2.5.1 Linguistic Domain Concepts (`LinguisticDC`)

The purpose of the linguistic subdomain (Fig. 2.5) is to define language related notions that can be refined for more specific domains. In particular, we want to distinguish natural languages spoken by human that can be part of a CPSs (e.g. human in the loop) from artificial languages such as those that can be processed by machines. We mostly follow definitions found in Wikipedia for these general notions.

2.5.1.1 Languages

According to Wikipedia a `Language` is "*a structured system of communication*". Since we want to distinguish between languages that are used by humans and those by machines or computers, this language class is then subdivided into the `NaturalLanguage` and `ArtificialLanguage` classes, where natural languages are those spoken by humans, such

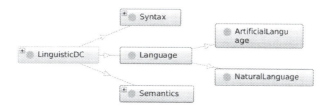

Figure 2.5 Overview of the `LinguisticDC` subdomain.

as English and French, while artificial languages are languages that can be processed by machines (computers). They are a simplification of natural languages.

2.5.1.2 Syntaxes

According to Wikipedia, a `Syntax` "*is the set of rules, principles, and processes that govern the structure of sentences (sentence structure) in a given language*". As a system of communication, every language (natural or artificial) must have at least one syntax to express sentences of the language to communicate with other entities. This is captured in our ontology with the `hasSyntax` object property, whose domain and range are, respectively, `Language` and `Syntax`. In the following, we will omit the specification of object property domains and ranges when obvious like in the case of the `hasSyntax` object property.

2.5.1.3 Semantics

According to Wikipedia, `Semantics` is the study of meaning in language, programming languages, formal logic, etc. It is concerned with the relationships between words, phrases, signs, and symbols and what they stand for in reality. In this ontology, we define semantics as the meaning of sentences of a language. As we will see in the MPM ontology, there are several ways to express meaning for a language by assigning semantics to sentences of a language. We will focus on those for the special case of modelling languages in the MPM ontology in Chapter 4.

Every language (natural or artificial) should have some meaning, in order for the language to be useful. This is captured in the ontology with the `hasSemantics` object property whose domain and range are respectively `Language` and `Semantics`. For artificial language, we are concerned that the meaning of a language is unique to avoid the ambiguity of natural languages. Therefore we define the `hasUniqueSemantics` object property as a subproperty of the `hasSemantics` property and we refine its domain to the `ArtificialLanguage` class. We further declare this property as being functional to indicate the uniqueness of semantics of artificial languages.

Many more notions related to linguistics could be added to this subdomain, the linguistic domain being quite vast and well studied. However, this is beyond the scope of this work and we therefore do not provide any other classes or properties at this time for the shared ontology.

2.5.2 Workflow Domain Concepts (`WorkflowDC`)

We introduce here core notions required to capture any workflow process. For this, we principally build on the workflow standard process definition language defined by the Workflow Management Coalition (WfMC) WFMC-TC-1025 standard [10]. It is to be noted that this standard also covers the business processes management domain as introduced in [11].

Fig. 2.6 displays the main process domain concepts taken from the WFMC-TC-1025 standard. For this framework, we only focus on the notions that will be needed to cover the more specific model-based development processes as defined in Chapter 5. In the following, we provide an overview of the concepts that we reuse / adapt from the standard and that are useful for this work, without giving all details of that part of the ontology. A complete description of the ontology can be found in [8].

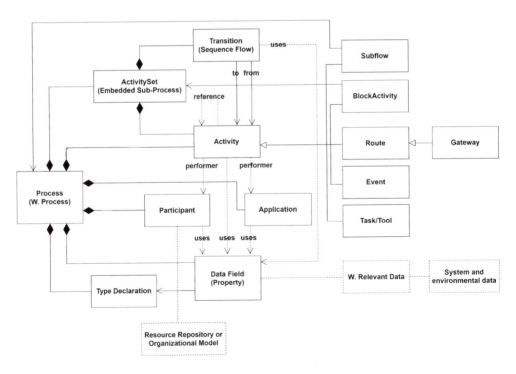

Figure 2.6 Workflow concepts (based on the WFMC-TC-1025 standard [10]).

Process

We define the `Process` OWL class (Fig. 2.7) as an entity providing contextual information applying to other entities within the process. Such information can be related to administration such as creation date and authors, or to process execution such as initiation parameters, execution priority, time limits, and person to be notified.

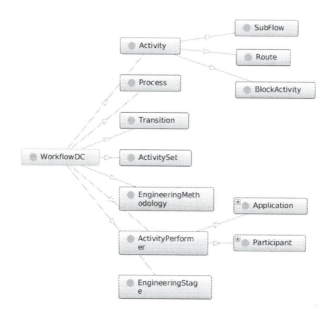

Figure 2.7 Overview of the `WorkflowDC` subdomain.

Activity

We define the `Activity` OWL class as "*a logical, self-contained unit of work within a process*". Such work will be performed by `ActivityPerformer` resources that we define as a subclass of the project `Resource` class from the project management subdomain. The `ActivityPerformer` class is further subclassed into the `Participant` class defined as "*a description of resources that can act as the performer of the various activities in the process definition*" and the `Application` class, defined as "*descriptions of the IT applications or interfaces which may be invoked by the service to support, or wholly automate, the processing associated with each activity*". We relate the `Activity` class to the `ActivityPerformer` class by defining the `hasActivityPerformer` object property with domain and ranges of the related classes.

The `Activity` class is subclassed into different kinds of activities. A `Block` activity consists of an embedded subprocess represented by an `ActivitySet`. A `hasActivitySet` object property is defined to link a `Block` to its `ActivitySet`. An `ActivitySet` contains a set of other activities and `Transition` elements declared via the `hasSetActivities` and `hasTransitions` object properties.

A `Transition` relates an activity via the *from* and *to* references to another activity to be executed if a condition defined by the transition is satisfied, or if no condition is set on the transition. This determines the execution order of the activities in the `ActivitySet`.

A `SubFlow` activity is a link to the execution of an independent subprocess declared in a process package (not shown in the figure). A `hasSubProcess` object property is defined to link a subflow activity to its process.

A `Route` is an activity that "*performs no work processing (and therefore has no associated resource or applications), but simply supports routing decisions among the incoming transitions and/or among the outgoing transitions*".

Processed data

The data that is manipulated by processes through activity performers such as `Application` and `Participant` is defined within a process definition through the *DataField* class of Fig. 2.6. This class will be replaced by the modelling notions of the MPM ontology and presented in Chapter 5.

Methodology

We extend the WFMC–TC–1025 standard to provide the concept of `Methodology` in order to be able to describe engineering processes such as that of the HPI CPSLab example at a higher level of abstraction. In science, according to Wikipedia, a methodology is defined as "*a general research strategy that outlines the way in which research is to be undertaken and, among other things, identifies the methods to be used in it*". However, a methodology does not define the specific employed methods but specifies the nature and kinds of processes to be followed in a particular procedure or to attain an objective.

We slightly update this definition to define an `EngineeringMethodology` as a general strategy that outlines the way in which engineering is to be undertaken and identifies a set of engineering stages defining the strategy. We therefore define an `EngineeringStage` class, which represents a categorisation of workflow processes to implement the stage of the methodology. And we define the corresponding `hasStages` object property. We also define the `hasNextStage` object property relating methodology stages in case the stages must be ordered.

A methodology is implemented by an overall workflow process defining activities supporting each stage of the methodology. So we define an object property `isImplementingMethodology` to relate a process to the methodology it implements. An additional `isImplementingStage` is defined to relate an activity to a stage it implements.

2.5.3 Project Management Domain Concepts (`ProjectManagementDC`)

The purpose of the project management subdomain is to define notions related to project management that can be relevant throughout the life cycle of systems. This domain is very rich and again it is not the purpose of the shared ontology to cover it entirely. However, high level project management notions such as stakeholders and organisations, project phases, and processes for supporting phases, are provided that can be used by other subdomains. We also found that defining these notions in the project management subdomain provides an adequate context to capture these notions, since CPS development always happens in the frame of projects.

The project management concepts of the shared ontology are principally inspired from the ISO 21500 standard [12]. We mainly focus here on the concepts related to stakeholders and their organisations, project, project phases and resources. This allows one to provide context for notions such as tools and development processes, which belong to resources of the project execution phase and that will be used by the MPM4CPS ontology of Chapter 5.

2.5.3.1 Stakeholders and project organisation

We mainly focus here on the stakeholder notion without considering the organisation that they necessarily belong to. Following the ISO 21500 standard [12], we define the `Stakeholder` OWL class (Fig. 2.8) as a *"person, group or organisation that has interests in, or can affect, be affected by, or perceive itself to be affected by, any aspect of the project."*

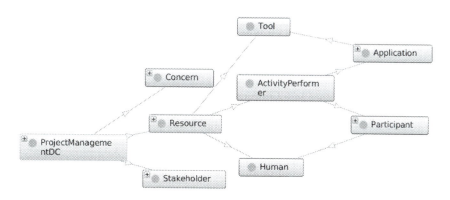

Figure 2.8 Overview of the `ProjectManagementDC` subdomain.

We also define the notion of stakeholder concern as the `Concern` class that we define as any topic of interest pertaining to a project or system to be developed during the project. We also provide the `hasConcern` object property to relate these two notions.

2.5.3.2 Resources

Resources are an important notion in project management as they are required to conduct project. They can often induce constraints on a project that must be taking into account such as their availability and costs.

Following the ISO 21500 standard, we define the `Resource` OWL class that we define as artifacts that are employed by projects. Resources can be humans, facilities, equipment, materials, infrastructure and tools. For this shared ontology, we are particularly interested in the human and tool kinds of resources. Therefore, we define the `Human` and `Tool` resources as subclasses of the `Resource` class. In addition, we want these resources to be usable in project processes that can be defined with classes of the workflow subdomain such as development processes. This subdomain provides the notions of `Participant` and `Application` activity performers. Therefore, we respectively make the `Human` and `Tool` resources subclasses of the `Participant` and `Application` classes. The `Tool` resource will be specialised into modelling tool in the MPM ontology.

2.5.3.3 Project and project phases

Typically, a project encompasses the phases of initiation, planning, execution and monitoring. Each phase makes use of processes for its execution. In this framework, we are mostly concerned with the execution phase within which system development happens. To capture how a system is being developed, we use the notions of the Workflow subdomain presented in Section 2.5.2 to specify processes of project phases.

2.5.4 Architecture Domain Concepts (`ArchitectureDC`)

The purpose of the architecture subdomain is to provide architecture-related concepts that are needed by both the CPS and MPM domains. As can be seen in Section 3.4, architecture is a structural property of real systems (and, consequently, of CPS). In addition, MPM also employs the notion of architecture, since it deals with the notions of architecture views, architecture models, architecture description languages, etc. Therefore, we introduce core architecture concepts in this shared ontology, which can then be augmented and refined with more specific notions pertaining to system structures in the CPS ontology and to architecture modelling in the MPM ontology.

Architecture frameworks have been developed to conceptualise a system architecture in order to assist understanding system essence and key properties pertaining to behaviour, composition and evolution, which in turn affect concerns such as system feasibility, utility and maintainability that must be taken into account by development projects [13]. Architecture descriptions improve communication and cooperation between stakeholders in order to enable them to work in an integrated, coherent fashion.

We borrow the notions of this architecture subdomain from the well-known ISO/IEEE 42010: Systems and Software Engineering - Architecture Description standard [14] (Fig. 2.9). However, in this shared ontology, which must remain independent of the CPS and MPM domains, we only retain the concepts that are not related to modelling since those will be detailed in the MPM and MPM4CPS ontologies of, respectively, Chapter 4 and Chapter 5. Therefore, we only introduce the top part notions of Fig. 2.9, excluding for now the notions of *ArchitectureViewpoint*, *ArchitectureView*, *ModelKind* and *ArchitectureModel*. Besides, we note that some concepts such as stakeholder and their concerns of Fig. 2.9 have already been introduced in the larger scope of project management of Section 2.5.3, which also describes the processes that elicit stakeholder and their concerns. Therefore, we reuse these notions for the architecture subdomain.

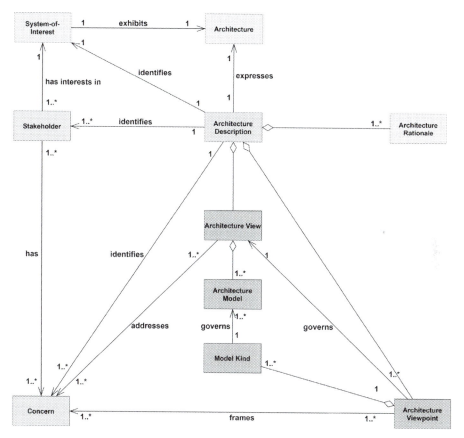

Figure 2.9 Concepts pertaining to the architecture description of a given system-of-interest (based on the IEEE 42010 standard [14]).

Following Wikipedia, we first define the System class (Fig. 2.10) as a group of interacting or interrelated entities that form a unified whole. Following the ISO/IEEE 42010, we then define the Architecture class as a property exhibiting the essence or fundamentals of a system. This definition is wide enough to encompass a variety of uses of the term architecture such as to understand and control those elements of a system-of-interest that contribute to its utility, cost, time and risk within its environment. We define the `hasArchitecture` object property to relate a system to its architecture.

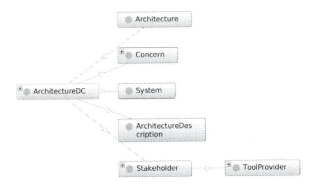

Figure 2.10 Overview of the `ArchitectureDC` subdomain.

We then introduce the `ArchitectureDescription` class as en entity that expresses one architecture for one system of interest. Those classes are related with the `hasIdentifiedSystem` and `hasExpressedArchitecture` object properties with corresponding domain and range classes. An architecture description also identifies stakeholders and their concerns via the `hasIdentifiedStakeholder` and `hasIdentifiedConcern` object properties. We reuse the stakeholder and concern classes of the project management subdomain for the range of these properties.

Other classes are not included in this subdomain of the shared ontology. As already mentioned, the model-related classes as well as the correspondence classes will be introduced in the MPM and MPM4CPS ontologies, while architecture rationale if it were needed for this work would be coming from the requirements subdomain.

2.5.5 Paradigm Domain Concepts (`ParadigmDC`)

The purpose of the paradigm subdomain is to provide paradigm notions that can be applicable to both the CPS and MPM domains and that can also be refined by them. The notion of paradigm is extremely vast and employed in several domains including CPS [15], programming [16,17] and obviously MPM as it is part of its acronym. In particular, as will be seen in Chapter 5, such notions will be used to characterise modelling paradigms.

In his book "The Structure of Scientific Revolutions" [18], the historian of science Thomas Kuhn defines a paradigm as "*a universally recognised scientific achievement that for a time provides model problems and solutions to a community of practitioners*". As further stated in Wikipedia, a scientific paradigm characterises the following aspects about the subject being based on the paradigm (candidate):

- what is to be observed and scrutinised;
- the kind of questions that are supposed to be asked and probed for answers in relation to the subject;
- how these questions are to be structured;
- what predictions made by the primary theory within the discipline;
- how the results of scientific investigations should be interpreted;
- how an experiment is to be conducted and what equipment is available to conduct the experiment.

In this work we are interested in a more restricted paradigm notion related to engineering. In [19], an interesting notion of engineering paradigm is presented following the work of Kuhn on scientific paradigms and paradigm shifts. Compared to scientific paradigms defined as "*characterising the intellectual environment in which science operates, made up of a set of accepted theories, assumptions, and methodologies under which ongoing work is carried out*", engineering paradigms are said to "*characterise the environment in which engineering takes place*". Like for science, engineering environments cover several aspects such as system design, design methods and tools and processes to develop systems. Examples of engineering paradigms are "*clean technology*" [20] that aims at minimising resource consumption and wastage during production processes and product life-cycle, or "*sustainable technological development*" [20] that concerns the development process by advocating that engineers should join public debates and closely interact with stakeholders (e.g., customers and politicians).

For scientific paradigms, a paradigm shift occurs when an existing paradigm (e.g. geocentrism in astronomy) can no longer explain all observations or experimental results. It is then replaced by a new paradigm (e.g. heliocentrism) that is able to explain all observations. According to [19], a major difference between engineering and scientific paradigm shifts is that engineering paradigm shifts are more predictable and change more gradually. New paradigms such as model-based engineering have emerged due to the inability of conventional approaches to address the development of increasingly complex systems. Also, new engineering paradigm may emerge as science evolves but also as new technologies are engineered and applied to the existing technology, requiring new paradigms in order to be able to apply the technology cost-effectively.

A commonality among the scientific and engineering paradigm definitions is that they are both a means to characterise a class of artifacts included in some environment. Therefore, in this shared ontology, we base our notion of engineering paradigm on the definition of [19]. In addition, some recent work has been published on the notion of *modelling paradigm* [21,22] as illustrated in Fig. 2.11. In this figure, we can see that classes are provided to also express a paradigm as a means to characterise some artifacts, which are specific to modelling and development process (expressed as workflows) in

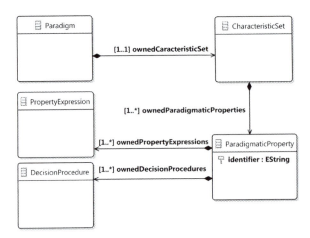

Figure 2.11 Concepts for modelling paradigms (based on [22]).

this case. Therefore, their definition is compatible with that of engineering paradigm being in fact a specialisation of engineering paradigm for engineering with models by characterising their formalisms.

Therefore, for the definition of engineering paradigms in this shared ontology, we can build of the generic notions of Fig. 2.11 that relate to paradigm characteristics. Other modelling-specific concepts will be introduced in Chapter 5 and based on the engineering paradigm notions of the shared ontology.

Therefore, we first define in the shared ontology the EngineeringEnvironment (Fig. 2.12) and EngineeringArtifact OWL classes. We give a broad definition of EngineeringArtifact as any artifact employed to engineer systems. The EngineeringEnvironment and EngineeringArtifact classes are related to each other via the hasArtifacts object property. It should be noted that engineering environments and their artifacts exist independently of any paradigm, a paradigm being a specific view of engineering environments.

We then define the EngineeringParadigm (*Paradigm* in Fig. 2.11) class and an object property isBasedOnParadigms to relate an engineering environment to the engineering paradigm it is based on. And we define the Characteristics (*CharacteristicSet* in Fig. 2.11) classes with an hasCharacteristics object property to relate these two classes.

Paradigm characteristics are defined by the ParadigmaticProperty class and related to Paradigm characteristics using the hasProperties object property. A paradigmatic property is expressed as a set of PropertyExpression associated with a paradigmatic property via the hasExpressions object property. Last but not least, it must be possible to decide if

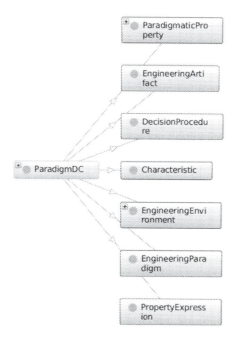

Figure 2.12 Overview of the `ParadigmDC` subdomain.

an artifact satisfies a paradigmatic property. This is achieved using a set of `DecisionPro-cedure` associated with a `ParadigmaticProperty` via the `hasDecisionProcedures` object property.

This completes the description of the paradigm subdomain of the shared ontology. Other notions to express modelling paradigms will be introduced in Chapter 5.

2.6. Introduction of examples

As explained in Section 2.2, the ontologies presented in the first four Chapters of this book have been developed via an exploratory modelling mode where existing CPSs and their modelling environments have been studied to derive the ontologies of Fig. 2.3. In doing so, two main examples of CPSs and their modelling environments have been used: the Ensemble-Based CPS development environment developed at Charles University in Prague and the *CPSLab*[15] of the Hasso Plattner Institute (HPI)[16] of the university of Potsdam. These two examples that will serve to illustrate the ontologies of Fig. 2.3 throughout Chapters 3, 4 and 5 are introduced in the following sections.

[15] http://www.cpslab.de.
[16] http://www.hpi.de.

2.6.1 Ensemble-based cyber-physical system

We present here a development environment for Ensemble-Based Cyber-Physical Systems (EBCPS) with respect to the ontologies of Fig. 2.3. We start with an overview of the features of EBCPS and its applicability in different domains. Then we present a CPS case study consisting of autonomous vehicle that illustrates the features of EBCPS followed by a description of models, languages and tools used for the development process. Later on, we will discuss in more detail the ontology perspective.

2.6.1.1 Overview

The EBCPS is a development environment for self-adaptive CPSs consisting of distributed *Components*, which form *Ensembles* under a specific context in the aim of collaboration. While the components contain the computational part and the interaction with the physical part, the ensembles represent the communication between the components and introduce the dynamism in the system. Such systems are general and have various applications in different domains such as traffic and transportation, robotics and clouds.

We use the DEECo computational model [23] (i.e. Dependable Ensembles of Emerging Components) to represent the main concepts to model EBCPS. DEECo is a realisation of the Service Component Ensemble Language (SCEL) (Fig. 2.13) introduced during the ASCENS project.[17] The DEECo concepts propose that a component should have knowledge (K) and processes (P), while the ensemble has a membership condition and exchanges knowledge between the components. However, the ensembles recognise the components in the group through their interfaces (i.e. roles – attributes as part of K), and perform the knowledge exchange if the membership condition evaluates to true. The membership condition over component attributes provides the system with context-awareness and dynamism.

Regarding the development process, we present a process and toolchain developed at the Department of Distributed and Dependable Systems[18] at Charles University in Prague (Fig. 2.14). The process starts with gathering the requirements and building the IRM-(SA) model. Afterwards, a transformation to Java code can be applied to implement the components and ensembles. Separately at design phase, the developer can design hierarchical structure for ensembles in Ensemble Definition Language (EDL) File using a plugin for jDEECo. Then it can transform the DSL to the corresponding Java code. In the code, it is also possible to support self-adaptation by using a provided set of Java annotations at the level of processes, where the annotation can represent a goal (i.e. from IRM-(SA)) or (a) mode(s). Additionally, the integrations with the simulators can be established through jDEECo at the code level using the corresponding plugins. After

---

[17] ASCENS project: http://www.ascens-ist.eu/.
[18] D3S: http://d3s.mff.cuni.cz/.

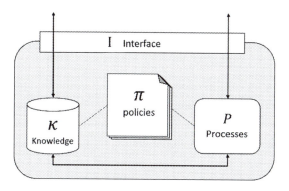

Figure 2.13 Service component (based on [24]).

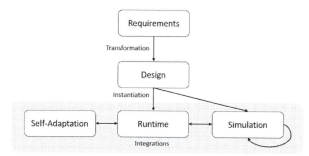

Figure 2.14 Overview of the development process of EBCPS using DEECo models and the operations between them.

finalising the code, the design model can be initiated, and the resulted runtime model executes the component processes and forms ensembles. At runtime, the self-adaptation, and the simulation run with DEECo runtime model.

2.6.1.2 CPS case study

The DEECo concepts used to model EBCPS can be applied in many domains since DEECo is not domain specific. However, some parts of the development process are domain specific, namely the simulation. The simulation tools used in DEECo were for robotics and for vehicles, thus we choose to present a case study in the transportation domain. This use case is a combination of several examples we published in the transportation domain with a small elaboration to provide a full representation of the development process and the toolchain use.

The case study contains autonomous vehicles that consider planning for parking during a trip. The vehicle travels from city A to city C through city B and the driver plans to stop in city B for sightseeing. During the trip, the vehicle joins a road train that has the same destination.

Joining a road train saves fuel and optimises the traffic in the highways. To join the train, the vehicle must communicate with the leader of the road train to ask for permission to join (Fig. 2.15). During the ride, the vehicles have to detect the distance from the vehicle in front to preserve a safe distance. Therefore, the vehicle detects the distance using camera, lidar and radar sensors in addition to having WiFi communication [25].

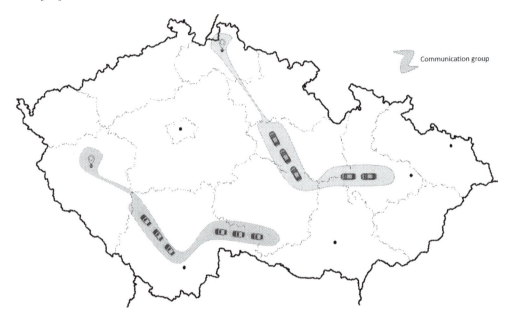

Communication group

Figure 2.15 Illustration of communication groups; each is associated with an instance of Same Destination (based on [26]).

When the vehicle almost reaches the city B, two options for parking inside the city are possible depending on the driver's needs. The driver might choose a single stop in the city or a few stops near to each other. In that case it is possible to reserve a place in the available parking lot nearest to the stop(s) [27]. If the driver chooses different stops scattered in the city, then more than one parking slot is required. However, if the driver does not have a strict time schedule for such scattered stops, the vehicle decides to communicate with nearby vehicles in the surrounding area and finds an available parking slot on the spot instead of reserving a place in a parking lot.

In the case of reserving a place in a parking lot, the availability of parking lots and the balance in the service [28] between parking lots should be taken into account to optimise the vehicle energy and the traffic flow in the city. In the case of finding a parking slot on the spot, the vehicle needs to communicate with other vehicles in the surrounding to know which vehicle is preparing for leaving soon or detected an available parking slot nearby [29] (Fig. 2.16). The vehicles can detect available parking slots using their cameras. This approach minimises the time spent to find a parking slot and saves energy in the process. However, boundaries on communication [30] are required to limit and optimise the broadcasting process of component knowledge for better performance, which is done by adding the boundaries on forming component ensembles [31].

As we can see, there are many scenarios associated to autonomous vehicles, and they require one to consider many issues such as adaptation (e.g. [32]) and communication problems.

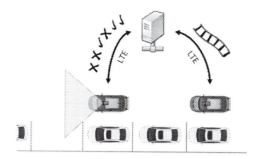

Figure 2.16 Scanning cars on the left side are sending a photo stream to the edge cloud server reporting spot availability to the parking cars on the right side (based on [29]).

2.6.1.3 Tool chain

For developing an EBCPS following the DEECo concepts, a toolchain is presented making use of custom made and existing tools (Fig. 2.17). Starting from requirements, the Invariant Refinement Method (IRM) [33] was developed using Epsilon,[19] where the developer can represent the invariants and assumptions in the system as a tree. The IRM tree ends with component processes and knowledge exchange in ensembles. At design phase, the implementation of the system can be done in Java (Fig. 2.18), C++ and Scala following the concepts of DEECo by using the developed runtime environment [23,34] jDEECo,[20] cDEECo,[21] or TCOOF[22] respectively. Most of the extensions and integration with other tools are done with jDEECo. As an example of the provided extensions, the self-adaptation plugin is an extension for jDEECo runtime in which the component modes and their transitions are represented as a state-machine. Regarding the integration, a number of simulation tools are integrated with the jDEECo runtime,

---

[19] https://www.eclipse.org/epsilon/.
[20] https://github.com/d3scomp/JDEECo.
[21] https://github.com/d3scomp/CDEECo.
[22] https://github.com/d3scomp/tcoef.

which are: MATSim for simulating vehicles [35], ROS for simulating robots [36] and OMNeT++ for simulating the delays in the network.

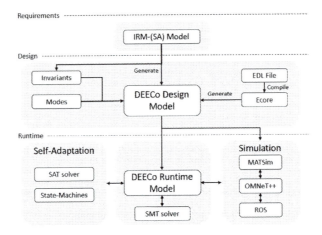

Figure 2.17 A toolchain for developing EBCPS.

```
@Component
public class Vehicle extends VehicleRole {
    ...

    @Process
    @PeriodicScheduling(period = 2000)
    public static void selectParkingProcess(...) {
        ...
    }
    ...
}


@Ensemble
@PeriodicScheduling(period = 2000)
public class CapacityExchangeEnsemble {

    @Membership
    public static boolean membership(...) {
        ...
        return false;
    }

    @KnowledgeExchange
    public static void map( ... ) {
        ...
    }
}
```

Figure 2.18 Snippet of jDEECo code (based on [35]).

2.6.2 HPI cyber-physical systems lab

In this section, we introduce the HPI CPSLab according to the different aspects covered by the ontologies of Fig. 2.3. We will present an overview of the lab including one of its cases study, its development process followed by the supporting toolchain and employed models and modelling languages. Details of each of these aspects will be presented in the corresponding Chapters 3, 4 and 5.

2.6.2.1 Overview

The HPI CPSLab applies an existing industrial-strength development methodology is-sued from the automotive domain [37] (Fig. 2.19), which has been adapted for the robotics domain [38]. It is based on MDE making use of architecture components decomposition supporting the combination of soft and hard real-time behaviour. It in-cludes several development activities such as modelling, simulation, verification/testing at different stages such as prototyping and pre-production.

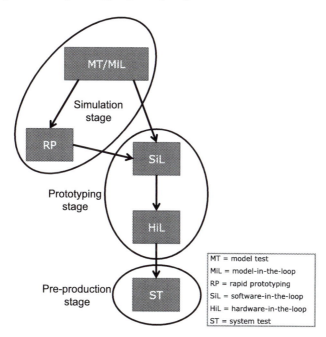

Figure 2.19 Overview of the HPI CPSLab development methodology.

2.6.2.2 CPS case study

The main CPS case study that is being addressed by the HPI CPSLab has also been used to develop the CPS ontology. It consists of a variable production setting where robots have the duty to transport pucks to different locations as advised by a factory automation (Fig. 2.20). The pucks represent goods of a real production system. The regular behaviour of a robot may be divided into three modes of operation consisting of moving around, transporting pucks and charging batteries. The robots must meet strict behaviour constraints such as avoiding complete discharge of the batteries, and with a lower priority, ensure to transport pucks as requested. The robots must also meet performance goals such as minimising energy consumption and maximising throughput. These can be achieved by short routing to the destination points while transporting the pucks. The robots must also avoid obstacles in hard real-time by reacting on obstacles within a few milliseconds.

Figure 2.20 Photo of the HPI CPSLab.

Fig. 2.21 depicts an overview of the overall robotics production CPS. It consists of four different rooms. In the first room, pucks are packed and dropped for transportation in area A_P. A robot R_P then transports a puck to a second room and drops it within the sorting area A_S. Based on the current delivery status, the robot R_S chooses one of the two booths and a band–conveyor transports the puck to the customer or stock delivery area (A_{CD}, A_{SD}). In a third step, the robot R_{St} transfers the puck to stock in St. The doors can be opened or closed dynamically to vary the scenario. A robot can charge its battery at one of the two charging points. Each robot is autonomous and therefore, the transportation, sorting and stocking tasks are independent from each other.

Three Robotino robots (Fig. 2.22) equipped with several sensors such as laser scanner, infrared (IR) distance sensors, GPS-like indoor navigation systems as well as different actuators such as servo motors, omnidirectional drive, and gripper are used in the production system.

While basic functionalities such as obstacle avoidance must be implemented in hard real-time, existing libraries are used to implement higher functionalities such as path

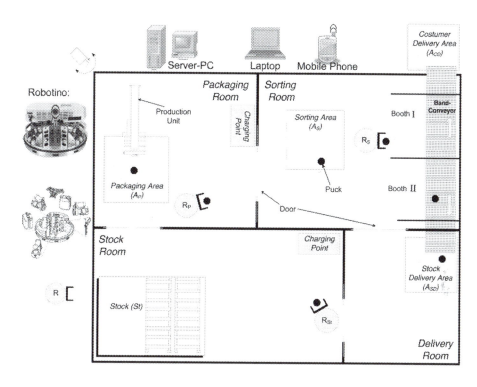

Figure 2.21 Structural overview of the employed evaluation scenario.

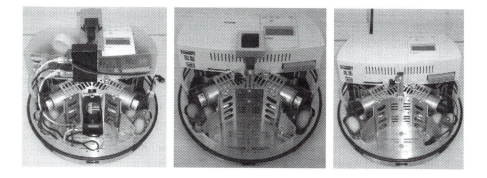

Figure 2.22 Photo of the employed robots.

planning or creating a map by evaluating measured distance values. The latter can rarely be performed under hard real-time constraints because of insufficient libraries. Furthermore, a RTAI Linux operating system is deployed on the robot to enable hard real-time execution.

2.6.2.3 Tool chain

In order to support the development process of the HPI CPSLab, several tools are combined including their related libraries in a toolchain that can handle the physical and cyber aspects of distributed robotics systems. In Fig. 2.23 the tool landscape is depicted. It consists of MATLAB®/Simulink for modelling and simulation, dSPACE SystemDesk for modelling software architecture, hardware configuration, and task mapping, dSPACE TargetLink for code generation and the FESTO Robotino-Library with the FESTO Robotino©SIM simulator.

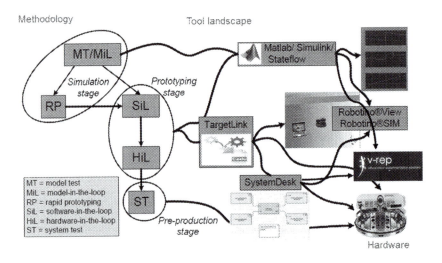

Figure 2.23 Tool landscape and its relation to the development methodology (from [38]).

2.7. Conclusion

This introduction chapter presented the motivation and context for the development of our ontological foundation for multi–paradigm modelling for cyber–physical systems, which is the subject covered by this first part of this book. We first introduced the approach used to develop the ontology following an exploratory modelling mode. We then presented the OWL and feature modelling languages that we used to support this modelling mode, including some ontology development best practices such as modularity and reuse. These practices led to the overall architecture of the framework that was then presented.

This included the definition of a *shared* ontology to capture notions that do not specifically pertain to the CPS and MPM domains but that are needed by these domains or that can be used to frame the more specific CPS and MPM notions. We introduced notions related to workflow processes, project management, architecture descriptions and engineering paradigms. Each of these notion builds on state of the art work such as standards. However, only the notions immediately required for this work were defined and completing these subdomains of shared ontology is left as future work.

Finally, we introduced the Ensemble-Based Cyber-Physical System and the HPI CPSLab development environments with their case studies that will be used to illustrate the ontologies throughout the next chapters.

References

[1] Thomas Kühne, Unifying explanatory and constructive modeling: towards removing the gulf between ontologies and conceptual models, in: Proceedings of the ACM/IEEE 19th International Conference on Model Driven Engineering Languages and Systems, MODELS '16, Association for Computing Machinery, New York, NY, USA, 2016, pp. 95–102.

[2] K. Kang, S. Cohen, J. Hess, W. Nowak, A. Spencer Peterson, Feature-oriented domain analysis (FODA) feasibility study, Technical Report CMU/SEI-90-TR-21, Software Engineering Institute, 1990.

[3] B. Tekinerdogan, K. Öztürk, Feature-Driven Design of SaaS Architectures, Springer London, London, 2013, pp. 189–212.

[4] K. Czarnecki, Chang Hwan, P. Kim, K.T. Kalleberg, Feature models are views on ontologies, in: 10th International Software Product Line Conference (SPLC'06), 2006, pp. 41–51.

[5] Thomas R. Gruber, Toward principles for the design of ontologies used for knowledge sharing?, International Journal of Human-Computer Studies 43 (5) (1995) 907–928.

[6] Hai H. Wang, Yuan Fang Li, Jing Sun, Hongyu Zhang, Jeff Pan, Verifying feature models using OWL, in: Software Engineering and the Semantic Web, Journal of Web Semantics 5 (2) (2007) 117–129.

[7] Rima Al-Ali, Moussa Amrani, Soumyadip Bandyopadhyay, Ankica Barisic, Fernando Barros, Dominique Blouin, Ferhat Erata, Holger Giese, Mauro Iacono, Stefan Klikovits, Eva Navarro, Patrizio Pelliccione, Kuldar Taveter, Bedir Tekinerdogan, Ken Vanherpen, COST IC1404 WG1 Deliverable WG1.2: Framework to Relate / Combine Modeling Languages and Techniques, Technical Report, 2020.

[8] Multi-paradigm modeling for cyber-physical systems website, http://mpm4cps.eu/, 2020.

[9] Stefan Klikovits, Rima Al-Ali, Moussa Amrani, Ankica Barisic, Fernando Barros, Dominique Blouin, Etienne Borde, Didier Buchs, Holger Giese, Miguel Goulao, Mauro Iacono, Florin Leon, Eva Navarro, Patrizio Pelliccione, Ken Vanherpen, COST IC1404 WG1 Deliverable WG1.1: State-of-the-art on Current Formalisms used in Cyber-Physical Systems Development, Technical Report, 2020.

[10] WFMC-TC-1025 Workflow Management Coalition Workflow Standard, Process Definition Interface – XML Process Definition Language, 2005.

[11] Mathias Weske, Business Process Management: Concepts, Languages, Architectures, third edition, Springer-Verlag, Berlin, Heidelberg, 2019.

[12] ISO 21500:2012: Guidance on Project Management, 2012.

[13] H.G. Gurbuz, B. Tekinerdogan, Analyzing systems engineering concerns in architecture frameworks – a survey study, in: 2018 IEEE International Systems Engineering Symposium (ISSE), 2018, pp. 1–8.

[14] ISO/IEC/IEEE 42010:2011. Systems and software engineering - architecture description, the latest edition of the original IEEE std 1471:2000, recommended practice for architectural description of software-intensive systems, 2011.

[15] Cees Lanting, Antonio Lionetto, Smart systems and cyber physical systems paradigms in an IoT and Industrie/y4.0 context, 2015, p. S5002.

[16] Brent Hailpern, Guest editor's introduction multiparadigm languages and environments, IEEE Software 3 (01) (jan 1986) 6–9.

[17] Pamela Zave, A compositional approach to multiparadigm programming, IEEE Software 6 (05) (sep 1989) 15–18.

[18] Thomas Kuhn, The Structure of Scientific Revolutions, Chicago Press, 2012.

[19] Erik W. Aslaksen, The engineering paradigm, International Journal of Engineering Studies 5 (02) (2013).

[20] Johannes Halbe, Jan Adamowski, Claudia Pahl-Wostl, The role of paradigms in engineering practice and education for sustainable development, Journal of Cleaner Production 106 (2015) 272–282, Bridges for a more sustainable future: Joining Environmental Management for Sustainable Universities (EMSU) and the European Roundtable for Sustainable Consumption and Production (ERSCP) conferences.

[21] M. Amrani, D. Blouin, R. Heinrich, A. Rensink, H. Vangheluwe, A. Wortmann, Towards a formal specification of multi-paradigm modelling, in: 2019 ACM/IEEE 22nd International Conference on Model Driven Engineering Languages and Systems Companion (MODELS-C), 2019, pp. 419–424.

[22] M. Amrani, D. Blouin, R. Heinrich, A. Rensink, H. Vangheluwe, A. Wortmann, Multi-paradigm modeling for cyber-physical systems: a descriptive framework, International Journal on Software and Systems Modeling (SoSyM), in press.

[23] Rima Al Ali, Tomas Bures, Ilias Gerostathopoulos, Petr Hnetynka, Jaroslav Keznikl, Michal Kit, Frantisek Plasil, DEECo: an ecosystem for cyber-physical systems, in: Companion Proceedings of the 36th International Conference on Software Engineering, ICSE Companion 2014, ACM, New York, NY, USA, 2014, pp. 610–611.

[24] Rocco De Nicola, Michele Loreti, Rosario Pugliese, Francesco Tiezzi, A formal approach to autonomic systems programming: the SCEL language, ACM Transactions on Autonomous and Adaptive Systems 9 (2) (2014) 7.

[25] Rima Al-Ali, Uncertainty-Aware Self-Adaptive Component Design in Cyber-Physical System, Technical Report D3S-TR-2019-02, Department of Distributed and Dependable Systems, Charles University, 2019.

[26] M. Kit, F. Plasil, V. Matena, T. Bures, O. Kovac, Employing domain knowledge for optimizing component communication, in: 2015 18th International ACM SIGSOFT Symposium on Component-Based Software Engineering (CBSE), May 2015, pp. 59–64.

[27] Nicklas Hoch, Henry-Paul Bensler, Dhaminda Abeywickrama, Tomáš Bureš, Ugo Montanari, The E-mobility Case Study, Springer International Publishing, Cham, 2015, pp. 513–533.

[28] Filip Krijt, Zbynek Jiracek, Tomas Bures, Petr Hnetynka, Frantisek Plasil, Automated dynamic formation of component ensembles - taking advantage of component cooperation locality, in: Proceedings of the 5th International Conference on Model-Driven Engineering and Software Development - vol. 1: MODELSWARD, INSTICC, SciTePress, 2017, pp. 561–568.

[29] Tomas Bures, Vladimir Matena, Raffaela Mirandola, Lorenzo Pagliari, Catia Trubiani, Performance modelling of smart cyber-physical systems, in: Companion of the 2018 ACM/SPEC International Conference on Performance Engineering, ICPE '18, ACM, New York, NY, USA, 2018, pp. 37–40.

[30] Tomas Bures, Ilias Gerostathopoulos, Petr Hnetynka, Jaroslav Keznikl, Michal Kit, Frantisek Plasil, Gossiping Components for Cyber-Physical Systems, Springer International Publishing, Cham, 2014, pp. 250–266.

[31] F. Krijt, Z. Jiracek, T. Bures, P. Hnetynka, I. Gerostathopoulos, Intelligent ensembles - a declarative group description language and java framework, in: 2017 IEEE/ACM 12th International Symposium on Software Engineering for Adaptive and Self-Managing Systems (SEAMS), May 2017, pp. 116–122.

[32] Rima Al Ali, Tomas Bures, Ilias Gerostathopoulos, Jaroslav Keznikl, Frantisek Plasil, Architecture adaptation based on belief inaccuracy estimation, in: 2014 IEEE/IFIP Conference on Software Architecture (WICSA), 2014, pp. 87–90.

[33] Jaroslav Keznikl, Tomas Bures, Frantisek Plasil, Ilias Gerostathopoulos, Petr Hnetynka, Nicklas Hoch, Design of ensemble-based component systems by invariant refinement, in: Proceedings of the 16th International ACM Sigsoft Symposium on Component-based Software Engineering, CBSE '13, ACM, New York, NY, USA, 2013, pp. 91–100.

[34] Tomas Bures, Ilias Gerostathopoulos, Petr Hnetynka, Jaroslav Keznikl, Michal Kit, Frantisek Plasil, DEECo: an ensemble-based component system, in: Proceedings of the 16th International ACM Sigsoft Symposium on Component-based Software Engineering, CBSE '13, ACM, New York, NY, USA, 2013, pp. 81–90.

[35] Michal Kit, Ilias Gerostathopoulos, Tomas Bures, Petr Hnetynka, Frantisek Plasil, An architecture framework for experimentations with self-adaptive cyber-physical systems, in: Proceedings of the 10th International Symposium on Software Engineering for Adaptive and Self-Managing Systems, SEAMS '15, IEEE Press, Piscataway, NJ, USA, 2015, pp. 93–96.

[36] Vladimir Matena, Tomas Bures, Ilias Gerostathopoulos, Petr Hnetynka, Model problem and testbed for experiments with adaptation in smart cyber-physical systems, in: Proceedings of the 11th International Symposium on Software Engineering for Adaptive and Self-Managing Systems, SEAMS '16, ACM, New York, NY, USA, 2016, pp. 82–88.

[37] Bart Broekman, Edwin Notenboom, Testing Embedded Software, Addison Wesley, 2003.

[38] Sebastian Wätzoldt, Stefan Neumann, Falk Benke, Holger Giese, Integrated software development for embedded robotic systems, in: Itsuki Noda, Noriaki Ando, Davide Brugali, James Kuffner (Eds.), Proceedings of the 3rd International Conference on Simulation, Modeling, and Programming for Autonomous Robots (SIMPAR), in: Lecture Notes in Computer Science, vol. 7628, Springer, Berlin, Heidelberg, October 2012, pp. 335–348.

CHAPTER 3

A feature-based ontology for cyber-physical systems

Bedir Tekinerdogan[a], Rakshit Mittal[b], Rima Al-Ali[c], Mauro Iacono[d], Eva Navarro[e], Soumyadip Bandyopadhyay[b], Ken Vanherpen[f] and Ankica Barišić[g]

[a]Wageningen University & Research, Wageningen, The Netherlands
[b]BITS Goa, Sancoale, India
[c]Charles University, Prague, Czech Republic
[d]University of Campania "Luigi Vanvitelli", Caserta, Italy
[e]University of Wolverhampton, Wolverhampton, United Kingdom
[f]University of Antwerp – Flanders Make, Antwerp, Belgium
[g]NOVA LINCS Research Center, Lisbon, Portugal

3.1. Introduction

Cyber-Physical Systems (CPSs) are systems that tightly integrate computation with networking and physical processes. Such systems form large networks that communicate with each other and rely on actuators and sensors to monitor and control complex with physical processes, creating complex feedback loops between the physical and the cyberworlds. CPSs bring innovation in terms of economic and societal impacts for various kinds of industries, creating entirely new markets and platforms for growth. CPSs have growing applications in various domains, including healthcare, transportation, precision agriculture, energy conservation, environmental control, avionics, critical infrastructure control (electric and nuclear power plants, water resources, and communications systems), high confidence medical devices and systems, traffic control and safety, advanced automotive systems, process control, distributed robotics (telepresence, telemedicine), manufacturing, and smart city engineering. The positive economic impact of any one of these applications areas is enormous.

Technically, CPS systems are inherently heterogeneous, typically comprising mechanical, hydraulic, material, electrical, electronic, and computational components. The engineering process of CPS requires distinct disciplines to be employed, resulting in a collection of models that are expressed using correspondingly distinct modelling formalisms. An important realisation is that distinct models need to be weaved together consistently to form a complete representation of a system that enables, among other global aspects, performance analysis, exhaustive simulation and verification, hardware in the loop simulation, determining best overall parameters of the system, prototyping, or implementation.

Multi-Paradigm Modelling Approaches for Cyber-Physical Systems
https://doi.org/10.1016/B978-0-12-819105-7.00008-8

A new framework is required that is able to represent these connections between models and, moreover, enable reasoning about them. No single formalism is able to model all aspects of a system; modelling of a CPS system is inherently multi-paradigm, which calls for a trans-disciplinary approach to be able to conjoin abstractions and models from different worlds. Physically, CPS systems are inherently heterogeneous, typically comprising mechanical, hydraulic, material, electrical, electronic, among others. Those areas correspond to engineering disciplines with their own models and abstractions designed to best capture the dynamics of physical processes (e.g., differential equations and stochastic processes). Computationally, CPS systems leverage the half-century old knowledge in computer science and software engineering to essentially capture how data is transformed into other useful data, abstracting away from core physical properties occurring in the real world, and particularly the passage of time in physical processes.

The key challenge, as identified a decade ago, is then to provide mathematical and technical foundations to conjoin physical abstractions that describe the dynamics of nature in various engineering domains, as described earlier, with models focusing solely on data transformation. This is necessary to adequately capture and bridge both aspects of a complex, realistic cyber-physical system, and become able to reason and explore system designs collaboratively, allocating responsibilities to software and physical elements, and analysing trade-offs between them.

This chapter aims to provide an ontology for CPS to support multi-paradigm modelling for CPS. To this end we will primarily use metamodelling and domain analysis. The result of the metamodelling activity will be a metamodel consisting of all the relevant concepts and their relations. For the domain analysis process we will adopt a feature model to represent both the common and the variant features of the CPS domain.

The remainder of this chapter is organised as follows. In Section 3.2 we present the overall approach for providing the ontology. Section 3.3 presents the metamodel and feature diagram for CPS as a result of the domain analysis process. Section 3.4 describes the CPS architecture that builds on the feature model and metamodel. Finally, Section 3.6 provides the conclusion.

3.2. Metamodel of cyber-physical systems

The domain analysis process has resulted in a set of concepts of CPS which have been represented in a metamodel as shown in Fig. 3.1 [1],[2]. The metamodel is a complementary model to a feature diagram and is used to show the relationships among the concepts. The feature diagram on the other hand stresses the commonality and variability of the CPS features.

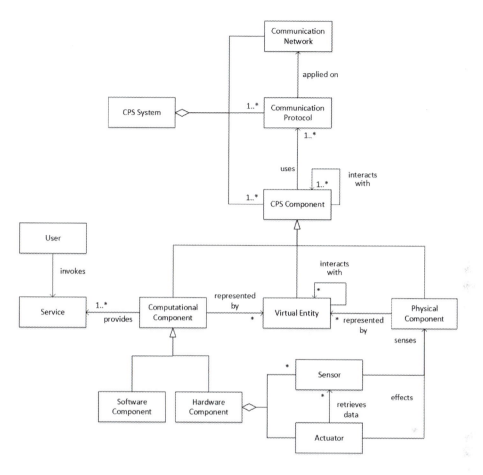

Figure 3.1 CPS metamodel.

3.3. Feature model of cyber-physical systems

To define the ontology for CPS we have carried out a domain analysis process using feature modelling. In the following sections we first describe the metamodel for CPS and then present the CPS feature diagram.

3.3.1 Top-level feature diagram

Fig. 3.2 shows the top-level feature diagram for CPS consisting of the five mandatory root features Constituent Element, Non-functional Requirements, Application Domain, Disciplines, and Architecture. The numbers indicate the number of sub-features of the corresponding root features. In the following sub-section we describe each root feature in detail.

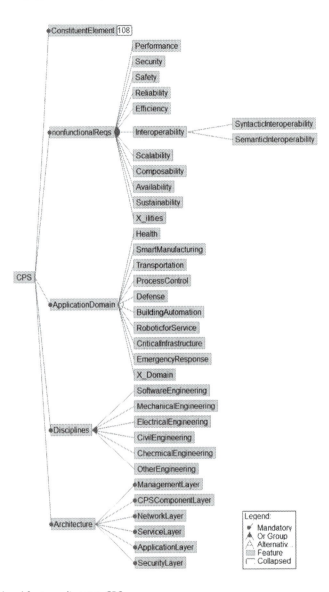

Figure 3.2 Top-level feature diagram CPS.

3.3.2 CPS constituent elements

Fig. 3.3 shows the feature diagram of the constituent elements of CPSs. Based on the feature diagram a CPS has the mandatory constituent elements of cyber element, physical element, control element, and network element. Further, human element can be an optional substituent element. Below, we describe each feature.

Cyber element

Communication and control between any human, biological, social system and any artificial device. It refers to systems where feedback is essential.

Physical element

The physical entity that is controlled by the cyber element, consisting of the sensor(s), actuator(s), plant, controller(s), and environment.

Control element

The action of modifying the behaviour of a system through feedback. The Control element is one of the most important and elaborate components of the CPS because it governs the interactions of the plant with its environment. Therefore, it becomes imperative to give a brief explanation of the Control element specifications i.e. its sub-features.

- **State**: The intrinsic configuration and description of the system.
- **Disturbance**: External influence of the environment in the system, typically unknown, but usually accounted for when designing the system.
- **Input**: Abstraction of the external factors in a system influencing its behaviour.
- **Output**: Abstraction of the effect of a system on its environment.
- **Goal**: The desired behaviour of a system. It can either be modelled by a set-point, which is a static goal (one that does not vary with time), or tracking the goal that varies with time. Another feature of the goal is its validity region – whether it is only valid in a subset of, or the complete state space of the system. The goal is reached by a regulation action. The goal is specified in implementations by the means of a reference signal, which models the desired value for the state of the system to reach.
- **Feedback**: The implementation of the control action by sensing the output/state of a system and modifying its input, by actuators, to meet a pre-defined control goal. The feedback is characterised by its dependency – whether the control action is designed depending on the system state or the system output. The scope of the feedback describes whether the feedback is applied to a centralised entity or not.
- **Dynamics**: This feature of the control element describes its evolution over time – its linearity, continuity, and time dependence (whether the system works in continuous or discrete time). Behaviour that does not exist in single systems, but when many systems come together and interact, the systems form a complex system that results in an emergent behaviour like swarming, flocking, collective, competition, etc. Such behaviour also comes under the domain of control element dynamics. The topology i.e. the structure of interconnections of the components of the CPS can also be either static or adaptive.
- **Properties**: Characteristics of the CPS related specifically to the control element, like stability, passivity, robustness, adaptation, controllability, autonomy, in-

telligence, consistency, learning and uncertainty. Uncertainty can be deterministic, non-deterministic, probabilistic, or stochastic.

- **Diagnostics**: Ability to identify the properties of the CPS.
- **Prognostics**: The ability to predict a time in the future, when the CPS will not perform as expected, anymore.

Network element

A set of elements (for example, nodes) connected in some physical or abstract manner (for example, links). A network can adopt different configurations such as star, bush, ring, mesh, point to point, and hybrid. Further, a network may use different communication mechanisms including the communication type (synchronous or asynchronous), and communication protocol (P2P, Client-Server, Broker, other).

Human element

Humans within a CPS perform certain roles within the CPS. Each role represents some capacity or position, where humans playing the role need to contribute for achieving certain behaviour goals set for the CPS. Each role is defined in terms of responsibilities and constraints pertaining to the role that are required for contributing to achieving the behaviour goals set for the CPS. Responsibilities are components of a role that determine what a human performing the role must do for the behaviour goals of the CPS to be achieved. Constraints are conditions that a human performing the role must take into consideration when exercising its responsibilities.

Humans within a CPS exercise their responsibilities defined by roles by performing certain actions. Action is an entity that is targeted at changing the state of the CPS or environment. Actions are divided into physical actions, communicative actions, and epistemic actions. Physical action is a kind of action that changes the state of a physical element of the CPS or environment. A communicative action is a kind of action that sends a message through a communication network of the CPS. An epistemic action is a kind of action that changes the state of the data held by the CPS. A human performs actions through actuators. A human can perceive events generated by the CPS or environment. An event is a kind of entity that is related to the states of affairs before and after it has occurred. A human perceives events through sensors. State of affairs is a collective state of the entities of the CPS and the environment. An entity is anything perceivable or conceivable.

3.3.3 Non-functional requirements

Cyber-Physical Systems revolutionise our interaction with the physical world. Of course, this revolution does not come free. Since even legacy embedded systems require higher standards than general-purpose computing, we need to pay special attention to

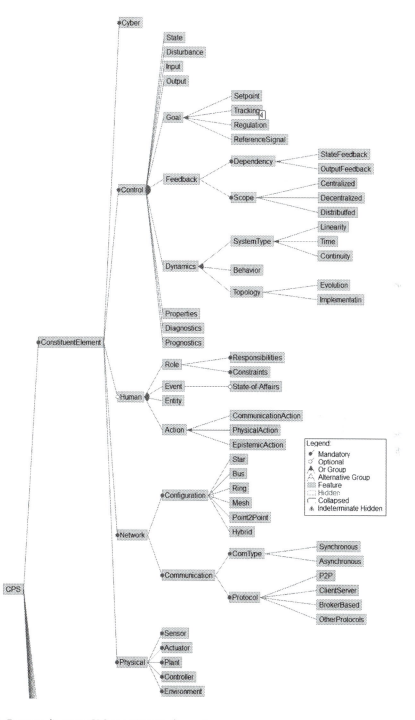

Figure 3.3 Feature diagram CPS constituent elements.

this next generation physically-aware engineered system requirements if we really want to put our full trust in them. Therefore, we want to clarify the definitions of some common CPS system-level requirements. In the top-level feature diagram of Fig. 3.2 the branch non-functional requirements include all the identified relevant CPS non-functional requirements.

Accuracy Accuracy refers to the degree of closeness of a system's measured/observed out-come to its actual/calculated one. A highly accurate system should converge to the actual outcome as close as possible. High accuracy especially comes into play for CPS applications where even small imprecisions are likely to cause system failures. For example, a motion-based object tracking system under the presence of imperfect sensor conditions may take untimely control action based on incorrect object position estimation, which in return leads to the system failure.

Adaptability Adaptability refers to the capability of a system to change its state to survive by adjusting its own configuration in response to different circumstances in the environment. A highly adaptable system should be quickly adaptable to evolving needs/circumstances. Adaptability is one of the key features in the next generation air transportation systems (e.g. NextGen). NextGen's capabilities enhance airspace performance with its computerised air transportation network which enables air vehicles immediately to accommodate themselves to evolving operational environment such as weather conditions, air vehicle routing and other pertinent flight trajectory patterns over satellites, air traffic congestion, and issues related to security.

Availability Availability refers to the property of a system to be ready for access even when faults occur. A highly available system should isolate malfunctioning portion from itself and continue to operate without it. Malicious cyber-attacks (e.g. denial of service attacks) hinder availability of the system services significantly. For example, in Cyber-Physical Medical Systems, medical data shed light on necessary actions to be taken in a timely manner to save a patient's life. Malicious attacks or system/component failure may cause services providing such data to become unavailable, hence, posing risk on the patient's life.

Composability Composability refers to the property of several components to be merged within a system and their inter-relationships. A highly composable system should allow re-combination of the system components repeatedly to satisfy specific system requirements. Composability should be examined in different levels (e.g. device composability, code composability, service composability, system composability). Certainly, system composability is more challenging, hence the need for well-defined composition methodologies that follow composition properties from the bottom up. Additionally, requirements and evaluations must be composable accordingly. In the future, it will probably be of paramount importance to incrementally add emerging systems to the system of systems (e.g. CPS) with some predictable confidence with-out degrading the operation of the resulting system.

Compositionality Compositionality refers to the property of how well a system can be understood entirely by examining every part of it. A highly compositional system should provide great insight about the whole from derived behaviours of its constituent parts/components. Achieving high compositionality in CPS design is very challenging especially due to the chaotic behaviour of constituent physical subsystems. Designing highly compositional CPS involves strong reasoning about the behaviour of all constituent cyber and physical subsystems/components and devising cyber-physical methodologies for assembling CPSs from individual cyber and physical components, while requiring precise property taxonomies, formal metrics and standard test benches for their evaluation, and well-defined mathematical models of the overall system and its constituents.

Confidentiality Confidentiality refers to the property of allowing only the authorised parties to access sensitive information generated within the system. A highly confidential system should employ the most secure methods of protection from unauthorised access, disclosure, or tampering. Data confidentiality is an important issue that needs to be satisfied in most CPS applications. For example, in an emergency management sensor network, attacks targeting confidentiality of data transmitted may degrade effectiveness of an emergency management system. Confidentiality of data transmitted through attacked sensor nodes can be compromised and that can cause data flow in the network to be directed over compromised sensors; critical data to be eavesdropped; or fake node identities to be generated in the network. Further, false/malicious data can be injected into the network over those fake nodes. There-fore, confidentiality of data circulation needs to be retained in a reasonable degree.

Dependability Dependability refers to the property of a system to perform required functionalities during its operation without significant degradation in its performance and out-come. Dependability reflects the degree of trust put in the whole system. A highly dependable system should operate properly without intrusion, deliver requested services as specified and not fail during its operation. The words dependability and trustworthiness are often used interchangeably. Assuring dependability before actual system operation is a very difficult task to achieve. For example, timing uncertainties regarding sensor readings and prompt actuation may degrade dependability and lead to unanticipated consequences. Cyber and physical components of the system are inherently interdependent and those underlying components might be dynamically interconnected during system operation, which, in return, renders dependability analysis very difficult. A common language to express dependability related information across constituent systems/underlying components should be introduced in the design stage.

Efficiency Efficiency refers to the amount of resources (such as energy, cost, time, etc.) the system requires to deliver specified functionalities. A highly efficient system should operate properly under optimum amount of system resources. Efficiency is especially important for energy management in CPS applications. For example, smart buildings can detect the absence of occupants and turn off HVAC (Heating, Ventilation, and Air Conditioning) units to save energy. Further, they can provide automated pre-heating or pre-cooling services based on the occupancy prediction techniques.

Heterogeneity Heterogeneity refers to the property of a system to incorporate a set of different types of interacting and interconnected components forming a complex whole. CPSs are inherently heterogeneous due to constituent physical dynamics, computational elements, control logic, and deployment of diverse communication technologies. Therefore, CPSs necessitate heterogeneous composition of all system components. For example, incorporating heterogeneous computing and communication capabilities, future medical devices are likely to be interconnected in increasingly complex open systems with a plug-and-play fashion, which makes a heterogeneous control network and closed loop control of interconnected devices crucial. Configuration of such devices may be highly dynamic depending on patient-specific medical considerations. Enabled by the science and emerging technologies, medical systems of the future are expected to provide situation-aware component autonomy, cooperative coordination, real-time guarantee, and heterogeneous personalised configurations far more capable and complex than today's.

Integrity Integrity refers to the property of a system to protect itself or information within it from unauthorised manipulation or modification to preserve correctness of the information. A high integrity system should provide extensive authorisation and consistency check mechanisms. High integrity is one of the important properties of a CPS. CPSs need to be developed with greater assurance by providing integrity check mechanisms on several occasions (such as data integrity of network packets, distinguishing malicious behaviours from the ambient noise, identifying false data injection and compromised sensor/actuator components, etc.). Properties of the physical and cyber processes should be well-understood and thus can be utilised to define required integrity assurance.

Interoperability Interoperability refers to the ability of the systems/components to work together, exchange information and use this information to provide specified services [3]. A highly interoperable system should provide or accept services conducive to effective communication and interoperation among system components. Performing far-reaching battlefield operations and having more interconnected and potentially joint-service combat systems, Unmanned Air Vehicles (UAVs) call for seamless communication between each other and numerous ground vehicles in operation. The lack of interoperability standards often causes reduction in the effectiveness of complicated and critical missions. Likewise, according to changing needs, dynamic standards should be developed and tested for devices, systems, and processes used in the Smart Grid to ensure and certify the interoperability of those ones being considered for a specific Smart Grid deployment under realistic operating conditions.

Maintainability Maintainability refers to the property of a system to be repaired in case a failure occurs. A highly maintainable system should be repaired in a simple and rapid manner at the minimum expenses of supporting resources, and free from causing additional faults during the maintenance process. With the close interaction among the system components (e.g. sensors, actuators, cyber components, and physical components) underlying CPS infrastructure, autonomous predictive /corrective diagnostic mechanisms can be proposed. Continuous monitoring and testing of the infrastructure can be performed through those mechanisms. The outcome of monitoring and testing facilities help finding which units need to be repaired. Some components, which happen to be the source of recurrent failures, can be redesigned or discarded and replaced with the ones with better quality.

Predictability Predictability refers to the degree of foreseeing of a system's state/behaviour/functionality either qualitatively or quantitatively. A highly predictable system should guarantee the specified outcome of the system's behaviour/functionality to a great extent every moment of time at which it is operating while meeting all system requirements. In Cyber-Physical Medical Systems (CPMS), smart medical devices together with sophisticated control technologies are supposed to be well adapted to the patient's conditions, predict the patient's movements, and change their characteristics based on context awareness within the surrounding environment. Many medical devices perform operations in real-time, satisfying different timing constraints and showing diverse sensitivity to timing uncertainties (e.g. delays, jitters, etc.). However, not all components of CPMS are time-predictable. Therefore, in addition to new programming and networking abstractions, new policies of resource allocation and scheduling should be developed to ensure predictable end-to-end timing constraints.

Reconfigurability Reconfigurability refers to the property of a system to change its configurations in case of failure or upon inner or outer requests. A highly reconfigurable system should be self-configurable, meaning able to fine-tune itself dynamically and coordinate the operation of its components at finer granularities. CPSs can be regarded as autonomously reconfigurable engineered systems. Remote monitoring and control mechanisms might be necessity in some CPS application scenarios such as international border monitoring, wildfire emergency management, gas pipeline monitoring, etc. Operational needs (e.g. security threat level updates, regular code updates, efficient energy management, etc.) may change for such scenarios, which calls for significant reconfiguration of sensor/actuator nodes being deployed or the entire net-work to provide the best possible service and use of resources.

Reliability Reliability refers to the degree of correctness which a system provides to perform its function. The certification of system capabilities about how to do things correctly does not mean that they are done correctly. So a highly reliable system makes sure that it does the things right. Considering the fact that CPSs are expected to operate reliably in open, evolving, and uncertain environments, uncertainty in the knowledge, attribute (e.g. timing), or outcome of a process in the CPS infrastructure makes it necessary to quantify uncertainties during the CPS design stage. That uncertainty analysis will yield to effective CPS reliability characterisation. Besides, accuracy of physical and cyber components, potential errors in design/control flow, cross-domain network connections in an ad hoc manner limit the CPS reliability.

Resilience Resilience refers to the ability of a system to persevere in its operation and delivery of services in an acceptable quality in case the system is exposed to any inner or outer difficulties (e.g. sudden defect, malfunctioning components, rising workload, etc.) that do not exceed its endurance limit. A highly resilient system should be self-healing and comprise early detection and fast recovery mechanisms against failures to continue to meet the demands for services. High resilience comes into play in delivering mission-critical services (e.g. automated brake control in vehicular CPS, air and oxygen flow control over an automated medical ventilator, etc.). Mission-critical CPS applications are often required to operate even in case of disruptions at any level of the system (e.g. hardware, software, network connections, or the underlying infrastructure). Therefore, designing highly resilient CPS requires thorough under-standing of potential failures and disruptions, the resilience properties of the pertinent application, and system evolution due to the dynamically changing nature of the operational environment.

Robustness Robustness refers to the ability of a system to keep its stable configuration and withstand any failures. A highly robust system should continue to operate in the presence of any failures without fundamental changes to its original configuration and prevent those failures from hindering or stopping its operation. In addition to failures, the presence of disturbances possibly arising from sensor noises, actuator inaccuracies, faulty communication channels, potential hardware errors or software bugs may degrade overall robustness of CPS. Lack of modelling integrated system dynamics (e.g. actual ambient conditions in which CPSs operate), evolved operational environment, or unforeseen events are other particular non-negligible factors, which might be unavoidable in the run-time, hence the need for robust CPS design.

Safety Safety refers to the property of a system to not cause any harm, hazard or risk in-side or outside of it during its operation. A very safe system should comply with both general and application specific safety regulations to a great extent and deploy safety assurance mechanisms in case something went wrong. For example, among the goals for Smart Manufacturing (SM), pointing-time tracking of sustainable production and real-time management of processes throughout the factory yield to improved safety. Safety of manufacturing plants can be highly optimised through automated process control using embedded control systems and data collection frame-works (including sensors) across the manufacturing enterprise. Smart networked sensors could detect operational failures/anomalies and help prevention of catastrophic incidents due to those failures/anomalies.

Scalability Scalability refers to the ability of a system to keep functioning well even in case of change in its size/increased workload, and take full advantage of it.

The increase in the system throughput should be proportional to the increase in the system re-sources. A highly scalable system should provide scatter and gather mechanisms for workload balancing and effective communication protocols to improve the performance. Depending on their scale, CPSs may comprise over thousands of embedded computers, sensors, and actuators that must work together effectively. Scalable em-bedded many-core architectures with a programmable interconnect network can be deployed to deliver increasing compute demand in CPS. Further, a high performance and highly scalable infrastructure is needed to allow the entities of CPS to join and leave the existing network dynamically. In the presence of frequent data dissemination among those entities, dynamic software updates (i.e. changing the computer program in run-time) can help update CPS applications dynamically and use CPS resources more productively.

Security Security refers to the property of a system to control access to the system re-sources and protect sensitive information from unauthorised disclosures. A highly secure system should provide protection mechanisms against unauthorised modification of information and unauthorised withholding of resources, and must be free from disclosure of sensitive information to a great extent. CPSs are vulnerable to failures and attacks on both the physical and the cyber sides, due to their scalability, complexity, and dynamic nature. Malicious attacks (e.g. eavesdropping, man-in-the-middle, denial of service, injecting fake sensor measurements or actuation requests, etc.) can be directed to the cyber infrastructure (e.g. data management layer, communication infrastructure and decision making mechanisms) or the physical components with the intent of disrupting the system in operation or stealing sensitive information. Making use of a large-scale network (such as the Internet), adopting insecure communication protocols, heavy use of legacy systems or rapid adoption of commercial off-the-shelf (COTS) technologies are other factors which make CPSs easily exposed to the security threats.

Sustainability Sustainability means being capable of enduring without compromising requirements of the system, while renewing the system's resources and using them efficiently. A highly sustainable system is a long lasting system which has self-healing and dynamic tuning capabilities under evolving circumstances. Sustainability from energy perspective is an important part of energy provision and management policies. For example, the Smart Grid facilitates energy distribution, management, and customisation from the perspective of customers or service providers by incorporating green sources of energy extracted from the physical environment. However, intermittent energy supply and unknown/ill-defined load characterisation hinders the efforts made to maintain long-term operation of the Smart Grid. To maintain sustainability, the Smart Grid requires planning and operation under uncertainties, use of real-time performance measurements, dynamic optimisation techniques for energy us-age, environment-aware duty cycling of computing units, and devising self-contained energy distribution facilities (such as autonomous micro grids).

3.3.4 Application domains

In the top-level feature diagram of Fig. 3.2 the branch Application Domain shows the important application domains for CPS. A CPS can be applied for various application domains including Health, Smart Manufacturing, Transportation, Process Control, Defence, Building Automation, Robotic Services, Critical Infrastructure, Emergence Response, etc. In principle the list is open ended. Any physical system that is integrated and controlled by a cyber part can be considered as CPS.

3.3.5 Disciplines

In the top-level feature diagram of Fig. 3.2 the branch Disciplines shows the important disciplines for CPS. In essence, CPS requires a holistic systems engineering approach. Systems engineering in its turn is an interdisciplinary approach that focuses on how to design, integrate, and manage complex systems over their life cycles. With this, CPS is inherently related to multiple disciplines including software engineering, mechanical engineering, electrical engineering, civil engineering, chemical engineering and others.

3.4. Architecture of CPS

The architecture of a CPS represents the gross level structure of the system consisting of cyber-physical components. Current architecture design approaches for CPS seem to be primarily domain-specific and no standard reference architecture has been yet agreed upon. In this line, the development of an ontology for CPS also contributes to the efforts made for designing a reference architecture.

A CPS reference architecture defines the generic structure of CPS architectures for particular application domains, laying the foundation for functionality, dependability, and other quality properties. An architecture organises the functionality and the properties of a system to enable partitioning, verification, and management. Fig. 3.4 presents a layered view of a CPS architecture inspired on the IoT stack that arranges a CPS into successive layers of cohesive modules that share similar concerns. The four layers at the centre include device layer, network layer, CPS component layer, application layer, and business layer. The CPS component layer includes the capabilities for the CPS components to undertake sensing and actuation. The network layer provides functionality for networking connectivity and transport capabilities enabling the coordination of components. The Services layer consists of functionality for generic support services (such as data processing or data storage), and specific support capabilities for the particular applications that may already apply a degree of intelligence. The application layer orchestrates the services to provide emergent properties. Then there are two main cross-cutting concerns. A Security layer captures the security functionality, while the management layer supports capabilities such as device management, traffic and congestion management.

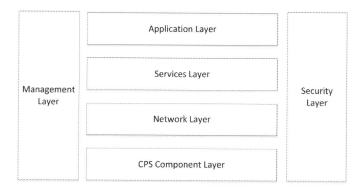

Figure 3.4 CPS architecture.

3.5. Examples

To demonstrate the application of the Cyber-Physical System concepts emulated in this chapter, the configuration of the CPS Feature Model (Section 3.3) pertaining to the examples described in Chapter 2 namely, Ensemble-Based Cyber-Physical System (Section 2.6.1) and HPI Cyber-Physical Systems Lab Autonomous Robot Case Study (Section 2.6.2) will be presented. A configuration is a particular instantiation of a Feature Model. It describes the features that are present or not present, in the particular product or system. Essentially, the Feature Model describes the inter-dependencies and constraints to be maintained in the features of each member of a family of systems and the configuration describes which features are present in the specific member. A configuration is deemed valid if it satisfies all the constraints of the parent Feature Model. The list of features, their inclusion or exclusion in the configuration, and the descriptions of the features pertaining to the example systems are given below:

3.5.1 Ensemble-based cyber-physical system

In order to present all the related parts in the example presented in Chapter 2, we refer to the tasks related to vehicle joining a road train as (1), and vehicle finding a parking slot as (2).

- ✓**ConstituentElement**: (mandatory) The elements constituting the system
  - ※ ✓**Cyber**: (mandatory) System model based on DEECo concepts.
  - ※ ✓**Control**: (mandatory) (1)(2) Vehicles have PID controllers to maintain the desired speed and the desired distance based on the driver preferences or the determined values in each vehicle mode in addition to (2) heading to the destination.
    - − ✓**State**: We determine the states, which are the operational modes, of (1) a vehicle in a road train to be "Cooperative Adaptive Cruise Control (CACC)" or "Adaptive Cruise Control (ACC)", and of (2) a vehicle parks

in a city to be "waiting", "search for parking lot", "reserve parking slot", "cancel reserved parking slot", "parking slot reserved", "search for parking slot on spot", "find parking slot on spot", "parking", "parked", or "leave the parking slot". The (2) parking lot has the following states for each parking slot: "available", "reserved", "cancelled", "filled".

— ✓**Disturbance**: (1)(2) Noise in sensors measurements, (1)(2) communication problems, and (1) traffic fluctuation.

— ✓**Input**: (1) In road train, the vehicle receives the speed and the position of the vehicle in front. (2) Regarding finding a parking slot, the input is the availability of the parking lots, the available slots detected by other vehicles,

— ✓**Output**: (2) In road train, the vehicle speed and distance from the vehicle in front is maintained. (2) When reaching the parking slot, the vehicle parks.

— ✓**Goal**: (1) Hard real-time goal keeps a save distance, (1) Soft real-time goal saves fuel, and (2) Soft real-time goal is parking.

  · ✓**Setpoint**: (1) The desired speed and the desired distance from the vehicle in front. (2) The final destination and the parking stops in between for the vehicle.

  · ✓**Tracking**: The vehicle checks for the driver's inputs for (1) the desired speed and (1)(2) the desired stops, or (1) updates from the vehicles in the road train for the speed and distance to maintain. (2) The vehicle receives information about available parking slots from other vehicles or from parking lots. The **ValidityRegion** of the tracking in (1) the vehicle is local when vehicle is not in a road train and decentralised when vehicle is in a road train. Additionally, (2) the vehicle tracks the available parking slots in decentralised manner. The **ValidityRegion** of the tracking in (2) the parking lot functions in decentralised manner.

  · ✓**Regulation**: (1)(2) The vehicles communicate with each other to maintain the road train or to find an available parking slots on the spot. (2) The vehicle communicates with the parking lot to reserve a parking slot.

  · ✓**Reference Signal**: (1)(2) The GPS signal is a reference signal in vehicles for self-positioning and (2) calculating the distance from the stops and parking lots.

— ✓**Feedback**: (1) There are two kinds of feedback loops in the autonomous vehicle. The first is part of the control loop to regulate the speed and the safety distance. We represent that loop using MAPE-K model [4]. The second feedback learns about the vehicle behaviour and the reliability of the sensors to perform better adaptation.

  · ✓**Dependency**: (1) The first feedback for the system is the measured acceleration from vehicle accelerometer (i.e. output of the Plant), which feeds back into the PID controller. The second feedback is studying the vehicle behaviour and the reliability of the sensors to decide adapting to a more suitable mode for the current situation (e.g. change from CACC to ACC in case the WiFi communication is unreliable).

· ✓**Scope**: (1) The scope of the control signals is local to the vehicle.

— ✓**Dynamics** Context-based ensembles i.e. The vehicles can form in dynamic groups to (1) maintain distance in a road train or (2) detect available parking slots, and (2) the vehicles form a dynamic grouping with the parking lots to exchange information about the parking slots.

· ✓**SystemType**: (**Linearity**) (1) The vehicle movement is non-linear. (**Time**) The system is discrete with relation to (1) the control loop in the vehicle and the communication with (1)(2) other vehicles or (2) the parking lots. (**Continuity**) The vehicle movement is continuous in reality. However, in the simulation it is discretized.

· **Behaviour** (**Equilibria**) There exist multiple **Distributed** equilibria for each car. The (**Emergent Behaviour)** of the CPS.

· **Topology** (**Evolution**) (1)(2) The connections between the components are dynamic over time (i.e. through ensembles), which (**Implementation**) support adding constraints over the connections that provide context awareness into the system.

— ✓**Properties**: Autonomy, adaptation, learning and uncertainty. (**Uncertainty**) In the example, we consider white noise in measurements and network delays (i.e. stochastic and exponential distribution accordingly).

— ✗**Diagnostics**: No automatic diagnostic.

— ✓**Prognostics**: The prediction is used in the system in the adaptation decision. For instance, after learning the vehicle behaviour, (1) the vehicle decides to change mode to ACC because the WiFi communication is unreliable.

∗ ✓**Human**: The driver determines the next tasks to be performed in the vehicle.

— ✓**Role**: (1) The driver decides to join or leave a road train, and (2) determines the final destination of the trip and the stops in between.

— ✓**Event**: (1) Detecting a road train, (2) Determine the locations to visit.

— **Entity**: The driver.

— ✓**Action**: the driver can make a (1) Request for joining or leaving a road train, or a (2) Request for finding a parking slot.

∗ ✓**Network**: (mandatory) The communication is Peer-to-Peer communication (i.e. MANET-based wireless and IP-based communication)

— ✓**Configuration**: (mandatory) The communication of peer-to-peer communication.

— ✓**Communication**: (mandatory) The communication is governed by protocols and constraints.

· ✓**ComType**: (mandatory) asynchronous communication since the communication is implicit, where the components propagate their knowledge

and do not wait for an answer. However, we assume a shared clock for all
the components in the example.

  · ✓**Protocol**: (mandatory) Gossipping via (2) UDP on top of Ethernet NIC
  and (1)(2) broadcast via wireless NIC.
* ✓**Physical**: (mandatory) The physical elements of the vehicle, the parking lots.
  − ✓**Sensor**: (mandatory) (1) When vehicle is on the road the used sensors are:
    WiFi antenna, GPS antenna, camera, radar, lidar, ultrasonic, and accelerom-
    eter. (2) When the vehicle is looking for parking the camera in the vehicles,
    and (2) camera or sensors in the parking lots to detect the availability of a
    parking slot.
  − ✓**Actuator**: (mandatory) In the vehicle (1) the gas and brake pedals (i.e.
    vehicle engine). Even though the publications did not cover the automatic
    parking; however, it is interesting to highlight the (2) steering as actuators in
    the vehicle.
  − ✓**Plant**: (mandatory) (1) The vehicle movement equations (because we sim-
    ulate)
  − ✓**Controller**: (mandatory) (1) The PID controllers
  − ✓**Environment**: (mandatory) (2) The cities, and (1)(2) the roads.
* ✓**nonfunctionalReqs**: Safety, Efficiency, Adaptability
* ✓**ApplicationDomain**: Transportation
* ✓**Disciplines**: The disciplines associated with the vehicle are − **Software Engi-
  neering** and **Mechanical Engineering**

3.5.2 HPI cyber-physical systems lab

* ✓**ConstituentElement**: (mandatory) The elements constituting the system
  * ✓**Cyber**: (mandatory) The cyber element of the system in the RTAI Linux OS.
  * ✓**Control**: (mandatory) The robot has a control system that moderates the phys-
    ical elements according to the different feedback inputs.
    − ✓**State**: The robots current position and orientation, and the task which
      it is performing (moving around, transporting pucks, or charging batteries)
      defines its state.
    − ✓**Disturbance**: Obstacle while performing task, or the reception of a new
      task which disturbs the current state of the robot (like battery level falling
      below a specified level)
    − ✓**Input**: Task to be performed.
    − ✓**Output**: Task performed (movement of pucks).
    − ✓**Goal**: Soft real-time goals are movement of pucks, ensuring battery is not
      depleted of charge. Hard real-time goal is avoidance of obstacles.
      · ✓**Setpoint**: The co-ordinates for the robot to reach.

- · ✓**Tracking**: The robot checks for new tasks or co-ordinates that may have been updated by the administrator. The **ValidityRegion** of the tracking is **local** to the robot. It only tracks for instructions concerning itself.
- · ✓**Regulation**: The actuators are activated in a co-ordinated manner to achieve the goals.
- · ✗**Reference Signal**: There is no reference signal because the robot does not know if its movement of the pucks is as desired i.e. successful.
- − ✓**Feedback**: There are various feedbacks for the control system.
  - · ✓**Dependency**: There are both, **StateFeedbacks** i.e. the feedbacks that describe the state of the robot, for example, the co-ordinates and orientation, the battery level and **OutputFeedbacks** i.e. describe the output of the robot like the feedback from the incremental encoder of the actuators that give information about the actuator's executed movements.
  - · ✓**Scope**: The scope of all the feedback signals is **centralised** to the control unit.
- − ✓**Dynamics** How the system behaves over time.
  - · ✓**SystemType**: (**Linearity**) The control system uses non-linear signal processing equations. (**Time**) The system is discrete because it functions w.r.t. edges of the clock. (**Continuity**) The control system is multi-modal continuous, but there is no direct interaction between the various dynamic systems of the robot (e.g. the omni-wheel drive, gripper, sensors, etc.) which is typical of such systems.
  - · ✓**Behaviour**: (**Equilibria**) There exists a **Common** equilibrium between all the different dynamic components, governed by the central control system. The (**Emergent Behaviour**) of the robot system is **Co-operation** as it ultimately cooperates with the various other robots and entities that may be present in the environment.
  - · ✓**Topology**: (**Evolution**) The interconnections between the different components of the robot stay the same over time, they are **Static**. (**Implementation**) The connections between the components are **Physical**.
- − ✓**Properties**: **Stability**, **Dissipativity** (energy in battery is dissipated over time), **Robustness** (object detection and path-planning), **Controllability**, **Observability**, **Resilience**, **Autonomy**, **Consistency**.
- − ✗**Diagnostics**: No automated diagnostic capability is present in the robot control system.
- − ✓**Prognostics**: Prognostics exist. That is why the indoor GPS–like architecture has been used so that the robot functions correctly according to the goal.

- ✓**Human**: The human operator defines the tasks for the robot.
  - − ✓**Role**: The administrator has the **Responsibility** of providing the co-ordinates for the robot to work with. The administrator does so with specific **Constraints** like reachability, correctness and syntax.
  - − ✓**Event**: Decision of the co-ordinates for the robot to pick-up and drop pucks.
  - − ✓**Entity**: The administrator.
  - − ✓**Action**: The human administrator performs a **Communication Action** where he/she communicates the required co-ordinates for the robot through a communication network.
- ✓**Network**: (mandatory) The network used in the system for different components to communicate with each other.
  - − ✓**Configuration**: (mandatory) The robot network has a star-type topology. All the components are connected to the central processing unit.
  - − ✓**Communication**: (mandatory) The communication is governed by protocols and constraints.
    - · ✓**ComType**: (mandatory) The communication is **Synchronous** since the sensors, actuators and the processing unit share the same clock.
    - · ✓**Protocol**: (mandatory) The network has some **Other Protocols** – half-duplex to interact with the various infrared distance sensors and laser scanners, whereas a full duplex protocol to interact with the administrator, and the different actuators.
- ✓**Physical**: (mandatory) The physical elements of the robot.
  - − ✓**Sensor**: (mandatory) Various sensors which provide feedback to the controller, like the infrared distance sensors, laser scanners, the incremental encoder from the drive unit of the actuators, the sensors comprising the indoor GPS-like navigation system, the communication antennae for communicating with the administrator, etc.
  - − ✓**Actuator**: (mandatory) The servo motors, omni-directional drive system, gripper, etc.
  - − ✓**Plant**: (mandatory)
  - − ✓**Controller**: (mandatory)
  - − ✓**Environment**: (mandatory) The HPI CPSLab environment where the robot operates.
- ✓**nonfunctionalReqs**: **Performance, Security, Safety, Reliability, Efficiency, Scalability, Composability, Availability, Sustainability, Others**.
- ✓**ApplicationDomain**: Domain of application of the robot can be broadly put under **RoboticforService** i.e. Service Robot.
- ✓**Disciplines**: The disciplines associated with the robot are – **Software Engineering, Mechanical Engineering, Electrical Engineering**.

3.6. Conclusion

In this chapter we have provided a feature-based ontology of cyber-physical systems. We have adopted feature modelling to represent the common and variant features of a CPS. The CPS feature model has been developed after a thorough domain analysis on CPS. Each feature branch and feature leaf has been carefully checked and described. The resulting feature model shows the configuration space for developing CPSs. We have used two different case studies on CPS and illustrated how to derive a concrete CPS configuration. Both case studies were applied after the feature model was designed but we were able to model all the features of the case studies, and did not need to adapt the CPS feature diagram. This was important to validate the external validity of the feature diagram. In our future work we will apply the CPS for other case studies as well.

The feature model is worthwhile for both researchers and practitioners. Researchers can identify the features of current CPSs and aim to identify novel features to enhance the CPS domain. The feature model can thus be used as a means to pave the way for further research in CPS. Practitioners can benefit from the resulting CPS by using it to understand and analyse existing systems and or develop novel CPSs. The CPS feature diagram has been derived based on a solid domain analysis. Further it has been validated by real world examples including Ensemble-Based Cyber-Physical System and HPI Cyber-Physical Systems Lab Autonomous Robot Case Study. As such, we can expect that the feature diagram is quite stable. However, in case of new development the feature diagram is adaptable and extensible to describe novel features.

References

[1] B. Tekinerdogan, Ö. Köksal, Pattern based integration of internet of things systems, in: Lecture Notes in Computer Science (including subseries Lecture Notes in Artificial Intelligence and Lecture Notes in Bioinformatics), vol. 10972, Springer Verlag, 2018, pp. 19–33.
[2] G. Giray, B. Tekinerdogan, E. Tüzün, IoT System Development Methods, CRC Press/Taylor & Francis, 2018, pp. 141–159.
[3] F. van den Berg, V. Garousi, B. Tekinerdogan, B.R. Haverkort, Designing cyber-physical systems with aDSL: a domain-specific language and tool support, in: 13th System of Systems Engineering Conference, SoSE 2018, Institute of Electrical and Electronics Engineers Inc., 2018, pp. 225–232.
[4] D. Sinreich, An architectural blueprint for autonomic computing, 2006.

CHAPTER 4

An ontology for multi-paradigm modelling

Holger Giese[a], Dominique Blouin[b], Rima Al-Ali[c], Hana Mkaouar[b], Soumyadip Bandyopadhyay[d], Mauro Iacono[e], Moussa Amrani[f], Stefan Klikovits[g] and Ferhat Erata[h]

[a]Hasso-Plattner-Institut, Potsdam, Germany
[b]Telecom Paris, Institut Polytechnique de Paris, Paris, France
[c]Charles University, Prague, Czech Republic
[d]BITS Goa, Sancoale, India
[e]University of Campania "Luigi Vanvitelli", Caserta, Italy
[f]University of Namur, Namur, Belgium
[g]University of Geneva, Carouge, Switzerland
[h]Yale University, New Haven, CT, United States

4.1. Introduction

In order to give a comprehensive presentation of modelling languages, modelling approaches, model composition, model construction, manipulation and execution, and model management, we developed an ontology-based description of the results of a wide study of the state of the art of modelling languages. The effort has been made to be as much inclusive as possible, with respect to what has been used or is fit for the description of the modelling aspects related to the reference domain, on the basis of the experience of the authors and of the participants of the COST action 1404 MPM4CPS – Multi-Paradigm Modelling for Cyber-Physical Systems. The main guideline of the investigation was based on previous results of Multi-Paradigm Modelling (MPM) work. This choice was made because of the comprehensive nature of the multi-paradigm approach, as it is designed to allow a systematic and semantically coordinated coexistence of submodels of different formal nature and expressive power in a single overall model. The ontology explores modelling concepts, linguistic aspects and formal subdomains of MPM including aspects related to formalisms, formal languages, formal semantics, modelling languages, model concepts, model relations (e.g. composition, decomposition, generation), modelling tools, general syntax aspects, specific languages, semantics, and modelling tools. Included concepts encompass metamodelling-based compositional approaches, model operations, flow composition and context composition approaches, global model management approaches (modelling-in-the-large) including mega-models, integration languages and multi-formalism modelling techniques.

In the rest of this chapter we first present a state of the art of modelling principles, concept and tools as well as multi-formalism and model management approaches to introduce the sources for our ontology. Being an enabler of MPM, besides the ontology, an important contribution of this chapter is our comprehensive state of the art treatment on model management that explores the numerous existing approaches and tools for model management.

We then present a description of the ontology and its analysis. The work is supported by some examples of MPM development environments for CPS that were used to derive the ontology and for demonstrating the coverage and the usefulness of the domain representation. The chapter is finally completed by some conclusions that propose our final considerations about this experience.

4.2. State of the art

Our MPM ontology aims at formalising the notions related to MPM by capturing them with OWL. This includes formalising the core notions related to modelling as well as the notions involved during the joint use of many formalisms employed when developing nowadays complex systems with MPM. This is also known as *multi-formalism modelling*. Together with metamodelling, multi-formalism allows the representation, the analysis and the synthesis of intricate knowledge for various domains and at various levels of abstraction [1]. As a result, numerous models must be jointly used to develop systems. Therefore these models must be managed properly to ensure that the models of different subsystems, of different views and of different domains are properly combined, even though the models might reside at different levels of abstraction. This is also known as *Global Model Management* (GMM) [2,3], for which a plethora of approaches has already been proposed.

Therefore, in this state of the art section, we present related work covering these three aspects. In Section 4.2.1, we first introduce existing work regarding the definition of core modelling concepts such as the notions of model, metamodel, and modelling language. Then we discuss multi-formalism modelling approaches in Section 4.2.2. Finally, we present a thorough state of the art on GMM concepts, approaches and tools in Section 4.2.3. Together, these notions form the basis of the ontology presented in Section 4.3.

4.2.1 Core modelling concepts

The proposed MPM ontology builds on definitions of concepts at the heart of modelling such as model, metamodel, formalism, languages, syntax, and semantics. Fortunately, we can build on some early research work defining these concepts such as [4] and [5]. In [5], agreed definitions for the core modelling concepts have been proposed by the MPM community. Fig. 4.1 illustrates these concepts represented as sets. In the figure, the inside ellipses represent subsets of the overall set of all models and labelled as "graph" since any model can always be represented as a graph. The dots inside the ellipses represent models (graphs) and the lines denote relationships between the models of the different kinds represented by the subsets. These subsets and their relationships therefore characterise the core concepts of abstract syntax, concrete syntax, metamodel, and semantic domain.

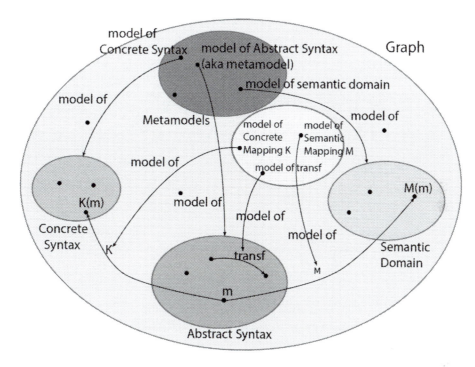

Figure 4.1 Modelling concepts as sets (from [5]).

The advantage of these definitions is that they were agreed upon by the MPM community and are therefore a perfect basis for developing our MPM ontology, with the advantage of representing these concepts formally with OWL instead of natural language. Besides, the representation of these concepts as subsets in Fig. 4.1 directly fits the semantics of OWL based on set theory.

Another very relevant work on the definition of core modelling concepts as well as other important MPM concepts is that of Broman et al. [6] characterising the notions of viewpoint, formalisms, languages and tools for the context of CPS development. The authors propose a framework consisting of viewpoints for capturing stakeholders' interests and concerns about the system, concrete languages and tools selected by stakeholders to address these concerns when designing a CPS, and abstract mathematical formalisms acting as the semantic glue linking the viewpoints to the concrete languages and tools. The framework is itself an adaptation of the ISO/IEEE standard 42010 [7] for the domain of CPS development.

Regarding core modelling concepts and compared to [5] these authors propose a slightly different notion of formalism and its relationship with languages as illustrated in Fig. 4.2. While in [5] a formalism is defined as being a language having a semantic domain and a semantic mapping function giving meaning to models of the language, Broman et al. define it as "mathematical objects consisting of an abstract syntax and a formal semantic". In their view, languages are concrete implementations of formalisms and may deviate slightly from the formalism in their implemented semantics. Besides, a language often implements multiple formalisms. As we will see in Section 5.3 and with the examples of Section 5.4, this definition is a good fit to reality.

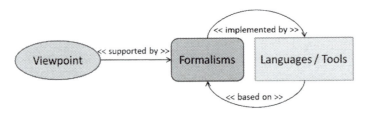

Figure 4.2 Relationships between viewpoints, formalisms, languages and tools (based on [6]).

4.2.2 Multi-formalism modelling approaches

Multi-formalism, Multi-resolution, Multi-scale Modelling (M4) environments may provide [8] an important and manageable resource to fulfil the modelling and simulation needs of modellers who have to deal with complex systems, where complexity derives from heterogeneity of components and relationships, multiple scales, multiple interacting requirements. Besides performance (or verification) oriented issues, multi-formalism approaches may also deal with software architecture oriented issues, e.g. by integrating UML as one of the formalisms to assist the development cycle of large, complex software systems [9]. In general, the literature proposes very popular dedicated transformational approaches for computer automated or assisted software generation that provide a formal framework to support the steps that lead from a formal or semiformal specification to code. However, in the rest of this subsection the focus lies on performance oriented approaches.

In multi-formalism modelling, many formalisms may be used simultaneously in a model. This may or may not exploit compositionality in the modelling approach, as elements of different formalisms may coexist in the model, the model may be composed of submodels written in different (single and particular) formalisms, or the different formalisms may be used in different steps of the processing of the model, by means of model transformation or generation. A general introduction to these themes can be found in [10].

Metamodelling is an important resource for both performance oriented approaches [11,12] and software-oriented transformation-based tools. Consequently metamodelling-based multi-formalism approaches can be considered a peculiar category. Another special category of approaches is constituted by the ones that deal with hybrid systems, that support multi-formalisms with both continuous and discrete nature, and that are thus capable of modelling natural systems in a better way [13]. These approaches should be able to describe and solve jointly and coherently differential equation-like descriptions and state space-based descriptions pertaining to the same complex system. While the problem has been popular in the 1970s and 1980s, there is currently a renovated interest from the point of view of cyber-physical systems: the interested reader can find specific general multi-formalism approaches in [14–16], that also provide an overview of selected previous, classical literature.

With reference to multi-formalism approaches oriented to performance evaluation, a number of different naive and structured approaches to the problem have been presented in the literature (a survey is provided in [17]). In the second group, the approaches have been implemented in a number of different tools, with different backgrounds, such as SHARPE, SMART, DEDS, AToM3, Möbius, OsMoSys and SIMTHESys. These tools also differ in the solution strategy adopted for the evaluation of models, and are designed with different purposes (e.g. some of them are designed to be extensible, some for experimenting with new formalism variants, some optimise the solution process).

SHARPE [18] supports the composition by submodels of some given different formalisms, solved by different solvers, but based on Markovian approaches. The composition consists of the exchange of probability distributions between submodels. SMART [19–21] supports the specification and solution, by simulation or approximation of complex discrete-state systems. DEDS [22] provides a common abstract notation in which submodels written in different formalisms are translated. Möbius [23–26] supports, by states and events superposition, a number of different formalisms (which can be extended by user-provided code) and alternative solvers (which can be chosen by the modeller) in a very articulated modelling and solution process.

Other approaches exploit, in different ways, metamodelling too. AToM3 [27,28] exploits metamodelling to implement model transformations, used to solve models by its solver. OsMoSys [29–33] and SIMTHESys [34–37] use metamodelling to let different user-defined formalisms interact by founding them over common metaformalisms and using elements and formalism level inheritance, and to implement different compositional mechanisms: while OsMoSys implements ad hoc operators for parameters exchange between submodels, and integrates external solvers by means of orchestration and adaptors, SIMTHESys privileges the experimentation of user-defined formalisms and embeds into formalism elements the interactions between different formalisms implementing multi-formalism by arcs superposition, allowing the automatic synthesis of proper solvers, according to the nature of the involved formalisms (with no claim for their optimality): there is an explicit specification of both syntax and semantics of every formalism element to allow high flexibility in the specification of custom, user-defined formalisms. For more details, the reader may refer to [10] providing a more detailed analysis on multi-formalism features and implementation, solution processes, purposes, compositional and transformational mechanisms of these approaches.

Most approaches are backed up with state space analysis techniques. Both analytical- and simulation-based methods are applied to perform the analysis, eventually with specific solutions to cope with the state space explosion problem, such as folding, decomposition, product forms solutions. The most common way is to directly generate the whole state space, with (e.g. in Möbius or in [38]) or without (e.g. in SMART or in some SIMTHESys solvers) an intermediate step of simple translation or more sophisticated transformation towards a specific intermediate representation, or by using partial state spaces exploiting modularity (e.g. in OsMoSys or in some SIMTHESys solvers [39]), or by transformation (e.g. in AToM, or in [40]). Noticeable are the approaches that exploit mean field analysis to cope with very large space states (e.g. [41] or [42]).

The literature provides a conspicuous number of applications of multi-formalism modelling: some significant examples are provided here. The effect of cyber-exploits on information sharing and task synchronisation has been studied in [43]; performance evaluation of Service Oriented Architecture has been studied in [44] and [45]; a cardiovascular system and its regulation have been studied with a hybrid approach, in [46]; interdependencies in electric power systems have been studied in [47]; the ERMTS/ETCS European standard for high speed trains has been studied in [48]; security attacks have been studied in [49]; exceptions–aware systems have been studied in [50]; effects of software rejuvenation techniques have been studied in [51]; NoSQL systems have been studied in [52]. Multi-formalism has been also applied as an implementation technique to provide higher-level tools or formalisms: in [53] a flexible, optimised Repairable Fault Tree modelling and solution approach is presented; a performance oriented model checking example is given in [54]; an analysis framework for detecting inconsistencies in high-level semantic relationships between models has been developed in [55].

4.2.3 Model management approaches

As already mentioned, the development activities for nowadays complex systems and in particular CPSs encompass multiple domains and teams, with each team using a dedicated set of modelling languages, thus requiring their proper integration and management. Using a single"model-it-al" language to cover all domains would certainly lead to large, monolithic languages that become less efficient, not easily customisable for development environments and tools needed by development teams, therefore adding difficulties to the already demanding effort of developing CPS. GMM approaches advocate the combination of reusable *modular* modelling languages instead of large monolithic languages. They support integrating models and *modelling languages* with appropriate abstractions and modularity, but also coordinating all activities operating on the models and specified as *model operations/transformations*. The execution of these model operations has to be *scalable* to handle large models. This requires *incrementality*, where only the operations impacted by a model change are re-executed, thus avoiding the effort to recompute entire models, as in the case of incremental code compilers.

Global Model Management (GMM) is also known as *modelling-in-the-large*, which consists of establishing global relationships (e.g. model operations that generated one model from other models) between macroscopic entities (models and metamodels), while ignoring the internal details of these entities [2]. Megamodelling [3,56] has been introduced for the purpose of describing these macroscopic entities and their relations. Nowadays only preliminary approaches exist that provide *ad hoc* solutions for fragments of the sketched problem and therefore, a solid understanding of the underlying needs including new foundations to address this problem is required as proposed to be developed by WG1 of MPM4CPS. In particular, the current approaches do at most offer some modularity and/or incrementality for a single aspect as modelling languages or model operations. However, support for handling complex modelling landscapes as a whole in a modular and incremental fashion as required for the large-scale problems that exist in practice is not offered so far.

In the following, we present existing solutions that address the features of GMM as depicted in Fig. 4.3. Section 4.2.3.1 presents solutions addressing the construction and execution of models and modelling languages achieved by means of links, model interfaces and metamodel composition. Section 4.2.3.2 describes solutions addressing the composition and execution of model operations / transformation according to flow (e.g. transformation chains) and context (e.g. transformation rules) composition approaches. Finally, Section 4.2.3.3 addresses the model management solutions making use of integration languages and megamodels.

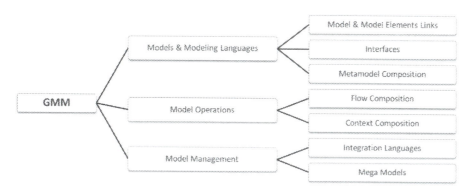

Figure 4.3 Features of global model management.

4.2.3.1 Construction of models and modelling languages

We discuss the construction of models and modelling languages using three main ways via (1) linking of model elements and models, (2) model interfaces and (3) metamodel composition.

Model elements and model links

Many approaches rely on *traceability links* between models and/or model elements to capture megamodelling relations/operations. We adopt here the definition proposed by the Center of Excellence for Software Traceability [57]: a *trace link* is *"a specified association between a pair of artifacts, one comprising the source artifact and one comprising the target artifact"*. The CoEST specialises those links into two dimensions: *vertical* trace links connect *"artifacts at different levels of abstraction so as to accommodate lifecycle-wide or end-to-end traceability, such as from requirements to code"*; while *horizontal* trace links associate *"artifacts at the same level of abstraction, such as: (i) traces between all the requirements created by 'Mary', (ii) traces between requirements that are concerned with the performance of the system, or (iii) traces between versions of a particular requirement at different moments in time"*.

A plethora of approaches has been proposed that make use of trace links for model integration (cf. e.g., [58–65]). The Atlas Model Weaving (AMW) language [58] provided one of the first approaches for capturing hierarchical traceability links between models and model elements. The purpose was to support activities such as automated navigation between elements of the linked models. In this approach, a generic core traceability language is made available and optionally extended to provide semantics specific to the metamodels of the models to be linked. Similarly, the Epsilon framework [59] provides a tool named ModeLink to establish correspondences between models. MegaL Explorer [60] supports relating heterogeneous software development artifacts using predefined relation types, linking elements that do not necessary have to be models or model elements. SmarfEMF [61] is another tool for linking models based on annotations of Ecore meta-models to specify simple relations between model elements through correspondence rules for attribute values. Complex relations are specified with ontologies relating the concepts of the linked languages. The whole set of combined models is converted into Prolog facts to support various activities such as navigation, consistency and user guidance when editing models. The CONSYSTENT tool and approach [62] make use of a similar idea. However, graph structures and pattern matching are used to represent the combined models in a common formalism and to identify and manage inconsistencies instead of Prolog facts as in the case of SmartEMF.

There are also a number of approaches, such as [63] and [64], that build on establishing links between models through the use of integration languages developed for a specific set of integrated modelling languages, where the integration language embeds constructs specific to the linked languages. This is also the case for model weaving languages extending the core AMW language. However, AMW has the advantage of capturing the linking domain with a core common language. Other means for linking and integrating models are Triple Graph Grammars (TGG) such as the Model Transformation Engine (MoTE) tool [65], which similarly requires the specification of some sort of integration language (correspondence metamodel) specific to the integrated languages. However, an important asset of this approach is that it automatically establishes and manages the traceability links and maintains the consistency of the linked models (model synchronisation) in a scalable manner. Finally, in [66–68], an approach is presented to automatically create and maintain traceability links between models in a scalable manner. While the approach focuses on traceability management rather than model integration, compared to integration languages, it relies on link types defined at the model level (and not at the metamodel / language level), thus avoiding the need to update the integration language every time a new language must be integrated.

The comparison of these approaches shows that apart from the approach [66–68], all approaches suffer from being dependent on the set of integrated languages, thus requiring to better support modularity. Furthermore, only [65–68] supports automated management of traceability links.

Interfaces

In addition to links, a few more sophisticated approaches (e.g., [69–71]) introduce the concept of *model interface* for specifying how models can be linked. In [69], the Analysis Constraints Optimisation Language (ACOL) is proposed, which has been designed to be pluggable to an Architecture Description Language (ADL). A concept of *interface* specific to ACOL is included so that constraints can refer to these interfaces to relate to the model elements expected from the ADL.

SmartEMF [70,72] proposes a more generic concept of model interface to track dependencies between models and metamodels and provides automated compatibility checks. Composite EMF Models [71,73] introduces *export* and *import* interfaces to specify which model elements of a main model (*body*) should be exposed to other models (i.e. are part of the public API), and which elements of a body model are to be required from an export interface.

MontiCore [74,75] also makes use of a notion of model interface of its own in order to compose grammars and corresponding models of independent languages. Furthermore, the approach makes use of feature modelling to capture modelling languages variability and commonalities to better reuse language components [76]. Finally, code generators can be composed to provide composed language semantics.

However, most of these approaches are only preliminary and need to be enriched to cover a larger number of model integration use cases such as for example, specifying modification policies of the linked model elements required to ensure the models can be kept consistent.

Metamodel composition

Some approaches (e.g., [77–81]) consider the construction of metamodels for expressing views in terms of other metamodels or language fragments. In [79], an approach is presented where metamodels are artificially extended for the purpose of combining independent model transformations resulting in an extended transformation for the extended metamodels.

The work in [82] presents operators to compose metamodels while preserving specific properties. In [83], the language and tool (Kompren) [77] is proposed to specify and generate slices of metamodels via the selection of classes and properties of an input metamodel. A reduced metamodel is then produced from the input metamodel. However, the produced metamodel must be completely regenerated when the input metamodel is changed. Such is the case for the Kompose approach [78], which on the contrary to Kompren, proposes to create *compound metamodels*, where a set of visible model elements from each combined metamodels is selected, and optionally related. The EMF Views [80,84] provides a similar approach, however, without the need to duplicate the metamodel elements as opposed to Kompose and Kompren where a new metamodel is created. These virtual view metamodels seem to be usable transparently by tools. Finally, the Global Model Management framework (GMM⋆)[1] [81] provides means to specify and interpret reusable language subsets as sets of constraints combined to form subsetted metamodels. Like for EMF Views, these reduced metamodels can to some extent be used transparently by tools. While each of these approaches provides interesting support for modular modelling languages, their unification into a common formalism, the use of an explicit notion of a model interface and their integration into GMM is lacking, except for subsetted metamodels already integrated within the GMM⋆ language.

The execution of integrated models concerns the evaluation of the well-formedness constraints of each combined model alone, but also of the combined models as a whole. To our knowledge, no approach addresses the incremental checking of well-formedness conditions across the different language fragments of compound models. However, some approaches on incremental constraints evaluation exist. In [85], changes on models are expressed as sequences of atomic model operations to determine which constraint is impacted by the changes, so that only these constraints need to be re-evaluated. In [86,87], a graph-based query language (EMF-IncQuery, which has later been renamed VQL and integrated into the VIATRA model transformation tool) relying on incremental pattern matching for improved performance is also proposed. In [88], an approach is presented for incremental evaluation of constraints based on a scope of model elements referenced by the query and determined during the first query evaluation. This scope is stored into

---

[1] We use ⋆ to distinguish this existing language and tool from the generic Global Model Management (GMM) acronym.

cache and used to determine which queries need to be re-evaluated according to some model changes. In [89], this approach is extended for the case where the constraints themselves may change besides the constrained models. Finally in [90], an incremental OCL checker is presented where a simpler OCL expression and reduced context elements set are computed from an OCL constraint and a given structural change event. Evaluating this simpler constraint for the reduced context is sufficient to assert the validity of the initial constraint and requires significantly less computation resources.

4.2.3.2 Construction of model operations

MPM approaches must support integrating models and *modelling languages* with appropriate abstractions and modularity, but also coordinating all activities operating on the models and specified as *model operations*. The execution of these model operations has to be *scalable* for being able to handle large models. This requires *incrementality*, where only the operations impacted by a model change are re-executed, thus avoiding the effort to recompute entire models, as in the case of incremental code compilers.

The construction of model operations is addressed in two ways in the literature. Most approaches combine model operations as *model transformations chains* (named (1) *flow composition*), where each chained transformation operates at the granularity of complete models. In order to support reuse and scalability for complex modelling languages, which are defined by composing them from simpler modelling languages, a few approaches have considered specifying model transformations as white boxes. Composed of explicit fine-grained operations processing model elements for a given context, these operations are reusable across several model transformations (named (2) *context composition*).

Flow composition approaches

Formal United System Engineering Development (FUSED) [64] is an integration language to specify complex relationships between models of different languages. It supports model transformation chains, but only implicitly via execution of tools, without explicit representation of the involved transformations and processed data. On the contrary, there is a plethora of approaches allowing the explicit specification and construction of model transformation chains implementing a data flow paradigm. A popular one is the AtlanMod Mega-Model Management (AM3) tool [91], for which the Atlas Transformation Language (ATL) [92] is used to specify the model transformations. Besides, a type system has been developed [93], which enables type checking and inference on artifacts related via model transformations. Another similar but less advanced tool is the Epsilon Framework [59], which provides model transformation chaining via ANT tasks. Wires [94] and ATL Flow [95] are tools providing graphical languages for the orchestration of ATL model transformations. The Formalism Transformation Graph + Process Model (FTG+PM) formalism [96] implemented in the AToMPM

(A Tool for Multi-Paradigm Modelling) tool [97] provides similar functionality. However, it has the advantage of also specifying the complete modelling process in addition to the involved model transformations. This is achieved via activity diagrams coupled with model transformation specifications executed automatically to support the development process. Finally, GMM★ [81] also supports model transformation chaining, but through the specification of relations between models of specific metamodels that can be chained. One advantage of this approach is that automated incremental (re-)execution of the specified relations between models is provided in response to received model change events. Incrementality of the execution of the transformations is also made possible by the integration of the MoTE [65] incremental model transformation tool into GMM★.

However, while chaining model transformations offers some degree of modularity of model transformation specifications, apart from GMM★, most approaches suffer from scalability issues for large models, since the used transformation tools do not support incremental execution. In addition, the case where a generated model is modified by hand to add information not expressible with the language of the original model(s) cannot easily be handled by these approaches, since regenerating the model modified by hand will destroy the user-specific information. This need is better supported by context composition approaches.

Context composition approaches

A few approaches allow context composition of model operations. In [79] as mentioned above, independent model transformations are combined, resulting in extended transformations for corresponding extended metamodels. In [98], the EMFTVM version of the ATL tool is provided where transformation rules can be combined at runtime to constitute complete model transformations. This allows reuse of rules used in the context of different transformations. In [99], view models are built using contextual composition of model operations (derivation rules) encoded as annotations of queries of the EMF IncQuery [86] language. Traceability links between view and source model elements are automatically established and maintained. The use of EMF IncQuery natively provides incremental execution of the derivation rules to synchronise the view model with the source model. Some views may be derived from other views, thus allowing flow composition as chains of view models. This approach achieves results similar to TGGs supporting incrementality, though with the drawback of being unidirectional. Similarly, but equipped with bi-directionality, the MoTCoF language [100] allows for both flow- and fine-grained context composition of model transformations. An advantage over [79], however, is that model transformations are used as black boxes without the need to adapt the transformations according to the context.

As can be seen, most approaches only support flow type modularity for model operations with batch execution except for the GMM★ language thanks to its integration of MoTE providing incremental execution. This will not scale and lead to information losses in case of partial model information overlap. Only a few approaches allow context modularity, which better supports incremental application where only the impacted operations can be re-applied following a change in order to avoid the cost of re-computing complete transformations. Such is the case of MoTCoF, which theoretically permits incremental execution, but a concrete technical solution is still lacking for it.

4.2.3.3 Model management approaches

Two strands can be identified for GMM. A first one makes use of (1) *model integration languages*, which are defined for a specific set of integrated modelling languages and tools, meaning that the integration language must be updated every time a new language or tool is used. The second strand attempts to solve this problem by making use of (2) *megamodels*, providing configurable model management.

Model integration languages

A classification of model integration problems and fundamental integration techniques has been introduced in [101]. It highlights the techniques of decomposition and enrichment, which characterise two orthogonal dimensions of development where the system is decomposed into subsystems and domains (*horizontal* dimension) and into a set of models with increasing level of details (*vertical* dimension).

The CyPhy [63] used in the Generic Modelling Environment (GME) modelling tool [102] and FUSED [64,103] are examples of model integration languages. But as mentioned above, these languages must be adapted as soon as a different set of integrated languages and tools must be used, thus requiring highly skilled developers. Integration languages are therefore not practical.

Open Services for Lifecycle Collaboration (OSLC) [104] provides standards for tool integration through the web. Many specifications are available for *change management*, *resource previews*, *linked data*, etc. It builds on the W3C *linked data* standard, which aims at providing best practices for publishing structured data on the web based on the W3C Resource Description Framework (RDF). RDF is a model for data interchange on the web where data is represented as graphs. However, OSLC is more services (and tools) oriented and inherits the problems of *linked data*, which is specific to the web and therefore does not separate the concerns of data representation and persistence, as opposed to Model-Driven Engineering (MDE) where an abstract syntax is used independently of the way the data is stored. Another approach making use of these standards is [62] and is implemented in the CONSYSTENT tool used to identify and resolve inconsistencies across viewpoints due to information overlapping. The information of all models involved during development is captured in a common RDF graph. The approach relies on a human (in parallel, an automated method making use of Bayesian Belief Networks is also under study [105]) to specify patterns representing semantic equivalence links (semantic connections) across the graph models. Inconsistency patterns based on these semantic connections are continuously checked over the RDF model for potential matches identifying inconsistencies. Means to automatically resolve inconsistencies are under development. However, since the conversion of all models as an RDF graph is required, this approach is not incremental and will not scale for large models.

Megamodels

In this second strand, megamodels serve to capture and manage MDE resources such as modelling languages, model transformations, model correspondences and tools used in modelling environments. There are several megamodelling approaches as already mentioned. AM3 [91] is one of the first initiatives where a megamodel is basically a registry for MDE resources. Model transformations are specified with ATL [92] and model correspondences with the Atlas Model Weaving (AMW) language [58]. Similarly, FTG+PM [96], as well as MegaL Explorer [60], allows one to model the artifacts used in software development environments and their relations from a linguistic point of view. The involved software languages, related technologies and technological spaces can be captured with linguistic relationships between them such as membership, subset, conformance, input, dependency, and definition. Operations between entities can also be captured. The artifacts do not need to be represented as models, but each entity of the megamodel can be linked to a web resource that can be browsed and examined. However, the language seems to be used mostly for visualisation providing a better understanding of the developments artifacts but cannot be executed to perform model management.

Model Management INTeractive (MMINT) [106,107] is an Eclipse workbench for graphical, interactive model management. MMINT provides a customisable environment, in which megamodels can be graphically created for model management scenarios at the instance level. In addition, MMINT provides a type level megamodel, that represents the collection of metamodels, relationships and transformations between them.

The MegaM@Rt2 toolbox [108] is composed of three complementary tool sets for System Engineering, Runtime Analysis and Model and Traceability Management, developed in the context of the MegaM@Rt^2 project [109], since 2017. MegaM@Rt2 Model and Traceability Management supports traceability across all layers of the system design and execution (runtime), by combining metamodelling and trace impact inference techniques. The toolbox makes use of NeoEMF for scalable model handling, EMF Views for model views and JTL for traceability management.

The aforementioned GMM★ infrastructure [81] consists of a mega-modelling language inspired from [110]. Metamodels can be declared, as well as relations between models of these metamodels. In particular, synchronisation relations can relate models of two different metamodels making use of the MoTE TGG engine [65] to transform or synchronise the models. As mentioned earlier, chains of model transformations can be specified and executed incrementally in response to model change events and *subsets* of modelling languages can be declared. GMM★ is experimented within the Kaolin tool [111] making use of complex and rich industrial languages such as AADL and VHDL thus challenging GMM for realistic specifications.

However, most of these megamodelling approaches only partially cover the core ingredients of specifying MDE resources by means of metamodels and model operations with appropriate modularity and incrementality. Only fragments of the problem are solved. Furthermore, all these megamodelling languages are monolithic and as a result, predefined megamodel fragments cannot be easily composed and reused to avoid rebuilding complete megamodel specifications from scratch for new projects. An attempt towards the reuse of megamodel fragments is presented in [112,113]. The work makes use of megamodelling techniques to propose an automated infrastructure to facilitate customisation, composition and reuse of the architect's representational resources to meet project, domain and organisation specific needs. Among these megamodelling approaches, only FTG+PM, GMM★ and [66,67] address the automated execution of megamodels in response to model changes or modelling events from the tool's user interface. GMM★ and [66,67] provide incremental execution of megamodels to some extent by re-evaluating only the relations concerned with the detected model changes.

4.3. MPM ontology

This section presents the MPM ontology based on the state of the art on MPM of the previous section and illustrated by the EBCPS and HPI CPSLab development examples of Section 4.4. The purpose of this section is not to present the entire ontology but only its main constituents as required by the examples. The complete specification of the ontology can be found in OWL form from our repository [114] or as a natural language specification generated from the OWL ontology in deliverables [115] and [116] of the MPM4CPS COST action.

Like for the shared ontology of Chapter 2, the MPM ontology provides a subdomain for notions related to modelling namely the ModelingDC as depicted in Fig. 4.4. The ModelingDC subdomain organises modelling concepts related to core modelling, multiformalism and model management as introduced in the state of the art of this chapter. It extends several notions introduced in a generic way by the shared ontology such as Language subclassed into ModelingLanguage and Tool subclassed into ModelingTool.

The modelling subdomain provides classes related to *scientific modelling* as defined in Wikipedia as "*a scientific activity, the aim of which is to make a particular part or feature of the world easier to understand, define, quantify, visualize, or simulate by referencing it to existing*

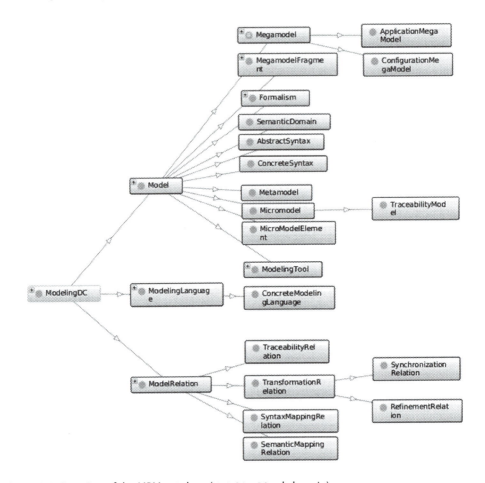

Figure 4.4 Overview of the MPM ontology (`ModelingDC` subdomain).

and usually commonly accepted knowledge".[2] This subdomain specifies the core notions of
`Model`, `ModelRelation`, `ModelingLanguage`, `AbstractSyntax`, `ConcreteSyntax`, etc., as well
as the core model management notions related to megamodelling such as `Megamodel`,
`MegamodelFragment`, and `ModelRelation`.

For this ontology, which is particularly concerned with modelling and model management, we have decided to reuse the megamodel notion as introduced in the state
of the art to support MPM. Megamodelling deals with "modelling in the large" in the
sense that it deals with a specific type of model whose elements are themselves models.
This therefore suggests two scales of modelling where we can consider usual models and

[2] https://en.wikipedia.org/wiki/Scientific_modelling.

their model elements (hereafter micromodels) and megamodels having micromodels as elements.

As shown in the state of the art, several definitions of megamodels have been proposed. In this ontology, we follow the unified definition of [110] as illustrated in Fig. 4.5. The advantage of this definition is that it can be extended to capture modelling notions at both the micro- and the mega-scales therefore unifying these two levels.

In the following, we first define OWL classes and object properties to follow the metamodel of Fig. 4.5. We then extend these notions to first define notions for the micromodelling scale. Then we provide another extension for the megamodelling scale.

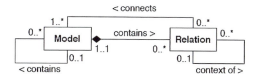

Figure 4.5 Unified core model definition (from [110]).

4.3.1 Core modelling notions

In its broadest sense, a model is an abstraction of reality that is defined by selecting and identifying relevant aspects of the real world, which are required for the activity(ies) to be performed with the model. Different kinds of models can be defined for different activities, such as conceptual models for understanding, mathematical models for analysis, and graphical models for visualisation.

It should be noted that while being a representation of reality, models can also be real objects themselves, such as a miniaturised building serving as a blueprint for the planned real-size building. However, for this ontology and following Fig. 4.1, we shall only consider models that are structurally represented as graphs, where model elements are the graph's nodes and relations between model elements are the graph's edges. Therefore we define an OWL class for representing models structured as graphs. Our Model class is equivalent to the set of all graphs of Fig. 4.1, given that in semantics of OWL, classes are sets as introduced in Section 2.3.2.

In Fig. 4.5, a model is hierarchical as it can contain other models via the *contains* reference. In this ontology, we rename this relation to define the hasContainedModels OWL object property. We adopt this hierarchical notion to reflect the actual structure of existing composite models such as UML. While a UML model is usually considered as a single monolithic model, conceptually it is composed of several sub-models following the separation of the UML language into its sub-languages such as class diagrams and use case diagrams.

Another essential notion in modelling is that of a ModelRelation. Following Fig. 4.5, a model relation is used to connect an arbitrary number of models. This *connects* reference is defined in our ontology by the `hasConnectedModels` object property. The property has the `ModelRelation` for its domain and the `Model` class for its range.

Note that a model relation is not considered to be a model in itself. This is to follow the underlying graph theory where a single edge is not considered to be a graph. A model relation only describes that models are related in some way depending on the type of the relation.

Besides, a model relation may exist in the context of another model relation. This is expressed by the *context of* reference of Fig. 4.5, which we translate as the `hasContextRelation` object property in the ontology. An example of such contextual relation may be a relation between to model elements that can only exist after a relation has first been created between their respective containing parent elements.

A model relation also needs to be contained in a model and therefore we create the `hasContainedModelRelations` object property whose domain is `Model` and range `ModelRelation`.

4.3.2 Micromodelling scale

We now provide classes and object properties for the notions of the micro-modelling scale by extending the aforementioned core modelling notions. We follow the example extension of Fig. 4.6 taken from [110]. In the figure, *M1Model* and *M2Model* correspond to the modelling levels of the OMG modelling pyramid defining metamodels with the conformance relation between a model its metamodel. Other classes of the figure will be introduced in the megamodelling scale section.

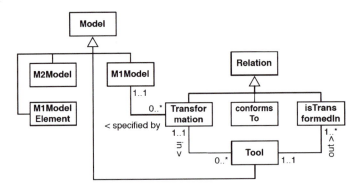

Figure 4.6 Example extension of the unified modelling metamodel (from [110]).

We first define a `MicroModel` as an instance-graph to represent models at the M1 level of the OMG modelling pyramid. The `MicroModel` class subclasses the core `Model` class.

We also define a `MicroModelElement` (corresponding to *M1ModelElement* in the figure) to represent atomic elements contained by a `MicroModel`. It subclasses the core `Model` class. We defined the `hasContainedModelElements` object property as a subproperty of the `hasContainedModels` object property with its domain and range, respectively, refined to `MicroModel` and `MicroModelElement`.

Thanks to the richness of OWL, we can further define the `MicroModel` class intensionally as a restriction stating that a `MicroModel` is a `Model` that only contains individuals of class `MicroModelElement` via the object property `hasContainedModels`.

4.3.3 Megamodelling scale

At the megamodelling level we define notions that relate to modelling in the large scale dealing with the representation of models of the micro-scale and relations between them. This includes notions related to modelling languages and the notions to model them such as metamodels and grammars. It also includes notions such as megamodel and megamodel fragment decomposing megamodels, as well as relations between models of the micro-scale such as transformations, synchronisation, and traceability.

4.3.3.1 Modelling language

According to Wikipedia, a modelling language is "*any artificial language that can be used to express information or knowledge or systems in a structure that is defined by a consistent set of rules. The rules are used for interpretation of the meaning of components in the structure.*" We refine this definition taking the definition of [5] stating that "*a language is the set of abstract syntax models*" (Fig. 4.1). We therefore first need to define the notion of abstract syntax.

Modelling languages syntaxes are traditionally partitioned into abstract and concrete syntaxes. An abstract syntax is a syntax that only captures the essence of a language (its concepts and the relations between them) independently from the notation users will employ to manipulate models of the language. An abstract syntax is equivalent to the Abstract Syntax Trees (ASTs) into which code of programming languages are transformed during compilation. A concrete syntax is the syntax users of the language manipulate to build models of the language. A modelling language may have several concrete syntaxes such as textual and graphical ones to improve its usability. A textual syntax consists of sequences of characters taken from an alphabet and grouped into words or tokens. Only some sequences of words or sentences are considered valid and the set of all valid sentences is said to make up the language. Graphical syntax display sentences of the language as visual elements such as boxes and arrows. However, pure graphical syntaxes are not common and often include textual parts.

We define the OWL `AbstractSyntax` class as a subclass of the `Syntax` class of the linguistic subdomain of the shared ontology. Being a subset of the model (graph) set of Fig. 4.1 we also make the `AbstractSyntax` class a subclass of the `Model` class.

We can now define a `ModelingLanguage` class as a subclass of the `Language` class of the linguistic subdomain. Since a modelling language is a set of abstract syntax models, it is therefore a subset of the graph set of Fig. 4.1 so we make the `ModelingLanguage` class a subclass of the `Model` class. We create the `hasAbstractSyntax` object property as a subproperty of the `hasSyntax` object property of the linguistic subdomain and we refine its `Language` domain and its `Syntax` range to, respectively, the `ModelingLanguage` and `AbstractSyntax` classes.

Similarly, we define the `ConcreteSyntax` class as a subclass of the `Syntax` class of the linguistic subdomain. Being also a subset of the model (graph) set of Fig. 4.1, we also make the `ConcreteSyntax` class a subclass of the `Model` class. In addition, since the two sets of abstract and concrete syntaxes are disjoint in Fig. 4.1 (an abstract syntax cannot be a concrete syntax), we also set these two classes as disjoint since OWL allows it.

As illustrated in Fig. 4.1, abstract and concrete syntaxes are related via a mapping function that associates element(s) of a concrete syntax to represent element(s) of the abstract syntax. We therefore define a `SyntaxMappingRelation` class as a subclass of the `ModelRelation` class and we create the `hasMappedAbstractSyntax` and `hasMappedConcreteSyntax` object properties to link the mapping relation to the mapped syntaxes. These object properties are made subproperties of the `hasConnectedModels` object property and their domains is refined to the `SyntaxMappingRelation` class. Their ranges are, respectively, refined to the `AbstractSyntax` and `ConcreteSyntax` classes.

In [5], the notion of *concrete language* is defined as a language that "*comprises both an abstract syntax and a concrete syntax mapping function k*". We therefore define a `ConcreteModelingLanguage` class as a subclass of the `ModelingLanguage` class. We define an object property named `hasSyntaxMapping` whose domain is `ConcreteModelingLanguage` and range is `SyntaxMappingRelation` to link a concrete language to its syntax mapping function.

According to Wikipedia, a *semantic domain* is defined as "*a specific place that shares a set of meanings, or a language that holds it*". Often a semantic domain can be made of another modelling language (abstract syntax) and its semantics. Such semantic domain together with a semantic mapping from models of the language to models the language of the semantic domain constitutes the semantics of the language.

We therefore define a `SemanticDomain` class as a subclass of the `Model` class. We also define a `SemanticMappingRelation` class as a subclass of the `ModelRelation` class and we create the `hasAbstractSyntaxSem` and `hasSemanticDomain` object properties to link the mapping relation to the mapped abstract syntax and semantic domain. These object properties refine the `hasConnectedModels` object property and their domains is refined to the `SemanticMappingRelation` class. Their ranges are, respectively, refined to the `AbstractSyntax` and `SemanticDomain` classes.

A semantic domain with a semantic mapping relation together constitutes the semantics of the language. This is also known as translational semantics. Therefore, we create a `TranslationalSemantics` class as a subclass of the `Semantics` class of the linguistic subdomain and provide an object property to relate the `TranslationalSemantics` class to its `SemanticMappingRelation`.

A metamodel (M2Model in Fig. 4.6) is a type-graph specifying the set of all valid models of a modelling language by specifying the language's abstract syntax. There must be a morphism between a model (instance-graph) and its metamodel (type-graph) for the model to be member of the language set. The expressiveness of metamodels is often not sufficient and a constraint language (such as the Object Constraint Language) must be used to specify additional constraints to further constrain the set of valid models.

We define the `Metamodel` class as a subclass of the `Model` class and a `ConformanceRelation` as a subclass of the `ModelRelation` between a `Metamodel` and its conforming `MicroModel`(s). Again, we create the `hasConformedModel` object property as a subproperty of the `hasConnectedModels` object property with its domain and range, respectively, refined to `ConformanceRelation` and `MicroModel`. We also create the `hasMetamodel` object property as a subproperty of the `hasConnectedModels` object property with its domain and range, respectively, refined to `ConformanceRelation` and `Metamodel`.

4.3.3.2 Megamodel

We now provide classes and object properties for the megamodelling notions implementing model management for MPM. Following [110], a *megamodel* is "*a model that contains models and relations between them. However, there is a major difference to classical software models that is a megamodel explicitly considers models instead of model elements*". We adopt this definition that covers several other megamodelling approaches for this MPM ontology.

Therefore, we create a `Megamodel` class as a subclass of the `Model` class. Being a model, a megamodel can contain `ModelRelation`(s) to relate models of the micro-scale. Thanks to the richness of OWL, we can easily express the second part of the definition as a restriction that checks that a megamodel does not contain any `MicroModelElements` via its `hasContainedModels` top level object property.

Besides, we introduce the `MegamodelFragment` class as a subclass of the `Model` class to represent parts of a megamodel. Being a model, like a megamodel, a megamodel fragment can contain `ModelRelation`(s) to relate models of the micro-scale. However, a megamodel fragment is not a megamodel in itself, hence its name. The fragment cannot be used alone as it requires other elements from other fragments to be usable. It should also be noted that megamodel fragments are just a logical organisation of megamodel elements and that they do not contain their elements, which are contained by the megamodel containing the megamodel fragment. We further introduce the object property `hasContainedMegamodelFragments` as a subproperty of the `hasContainedModels` object property with, respectively, the `Megamodel` and `MegamodelFragment` classes for the property's domain and range.

We now introduce more specific megamodelling relations useful to capture MPM development environments. A transformation relation is a relation that can be executed to transform some source model(s) into some target model(s). Such relation can be bidirectional when the source models can play the role of target models and vice versa. It can be specified by a transformation specification represented as a model of a transformation language (Fig. 4.6).

Therefore our ontology provides a `TransformationRelation` class as a subclass of the `ModelRelation` class with an object property `hasSpecification` to relate the transformation relation to its specification. The domain and range of the property are, respectively, the `TransformationRelation` and `MicroModel` classes. In addition, we define the `hasInputModels` and `hasOutputModels` object properties whose domains and ranges are, respectively, `TransformationRelation` and `ModelingLanguage`.

After or while a transformation is being executed, the transformed models often need to be related and a set of traceability relations are created as a by-product of the transformation execution. As already mentioned, such relations may often only exist in the context of other relations. Besides, often all these relations are grouped under a traceability model. This is the case for transformation tools based on Triple Graph Grammars (TGG) where a third traceability model is constructed as required for model synchronisation.

In order to model this in our MPM ontology, we create a `TraceabilityRelation` class as a subclass of `ModelRelation`. We also create a `TraceabilityModel` class as a subclass of `MicroModel` and we refine the `hasContainedModelRelations` object property into the `hasTraceabilityRelation` with the domain and the range, respectively, refined into `TraceabilityModel` and `TraceabilityRelation`.

Different kinds of transformation relation exist that can be modelled in our MPM ontology. A `SynchronizationRelation` is a subclass of `TransformationRelation` where only parts of the models that need to be updated to preserve consistency are changed instead of generating complete models when the transformation is executed.

We also modelled other more specific transformation relations such as a refinement relation where the level of abstraction of the target model is lower than that of the source model. Such relations are frequently used as steps towards code generation such as in the RAMSES tool.[3]

4.3.3.3 Activity performers

The shared ontology of Chapter 2 defines the `Human` and `Tool` classes to represent resources that can perform activities of workflow processes. Those classes are also subclasses of the `ActivityPerformer` class. Even when not made explicit in a tool or a human performer, both these performers make use of transformation relations to specify their possible model manipulations. Therefore we define the `hasTranformationSpecifications` object property to relate an activity performer resource to the transformation relations it executes (Fig. 4.6).

---

[3] https://mem4csd.telecom-paristech.fr/blog/index.php/ramses/.

Finally, we subclass the `Tool` class of the project management subdomain of the shared ontology to define the notion of `ModelingTool`. We also define a `ModelingHuman` class as a subclass of the `Human` class of the project management subdomain to represent human performing modelling activities. Like for the *Tool* class of Fig. 4.6, both these classes are made subclasses of the `Model` class.

We also create the `isToolFor` object property to relate a modelling tool to its employed modelling languages. The values of this property can be derived as being the languages connected by the tool's transformation relations.

We also provide other subclasses of `ModelingTool` for more specific modelling-related activities such as `SimulationTool`, `TransformationTool`, `VisualizationTool`, and `ExecutionTool`. Detailed definitions of these tools can be found from the complete ontology specification [114].

4.3.3.4 Formalism

In [5], a *formalism* is defined as "*a language, a semantic domain and a semantic mapping function giving meaning to model in the language*". In the framework of Broman et al. [6], formalisms are "*mathematical objects consisting of an abstract syntax and a formal semantics*" and modelling languages are "*concrete implementations of formalisms*". The authors further note that a language's semantics may slightly deviate from the semantics of the formalisms they implement. In addition, it is often the case that a language implements several formalisms. For instance, the AADL language [115] is based on several formalisms including entity-relationships for its components decomposition, state machines for its mode construct, behaviour annex and error model annex sub-languages, and data flow for its data port notion. As a matter of fact, the formalism notion of Broman et al. is close to the notion of *modelling paradigm* as we will discuss in further detail in Chapter 5.

The relationships between formalisms, modelling languages and tools are illustrated in Fig. 4.2. In order to capture these notions in our MPM ontology, we first define a `Formalism` class making it a subclass of the `Model` class. We then create an `isImplementedBy` object property with, respectively, the `Formalism` and `ModelingLanguage` classes as domain and range. We also create a `isBasedOn` object property with, respectively, the `ModelingLanguage` and `Formalism` classes as domain and range and we make this property the inverse of the `isImplementedBy` object property as allowed in OWL.

Many formalisms such as Petri nets or automata have different variants. For example, for Petri nets we have High Level Petri nets, Petri nets with Priority, Stochastic Petri nets, all of which can be implemented by the PNML language.

Therefore, we define subclasses of the `Formalism` class such as `PetriNetBasedFormalism`, `AutomataBasedFormalism`, `LogicBasedFormalism` to reflect this finer grained formalisms classification.

4.4. Examples

In this section, we use the MPM ontology introduced in Section 4.3 to capture the details of the different CPS development environments examples introduced in Chapter 2. This includes defining the employed modelling artifacts such as models, modelling languages, modelling tools, formalism, and model relations. This is performed for each step of the development processes of the two examples in order to validate the MPM ontology.

4.4.1 Ensemble-based cyber-physical system
4.4.1.1 Overview

The Ensemble-Based Cyber-Physical Systems (EBCPS) development environment uses a number of models to support the activities performed at each stage of its development process. Such process is depicted in Fig. 2.14. The tools used for development are the jDEECo runtime and its plugins. The plugins include tools for capturing requirements, forming ensembles, supporting self-adaptation mechanisms at runtime, and supporting co-simulation with MATSim and ROS. The simulator tool MATSim is used to simulate vehicles movements, while ROS is used for simulating robots.

We use the MPM ontology to represent the elements of EBCPS in Fig. 4.7, and present an overview of the megamodels, model transformations and tools used at the different development stages of the CPS use case introduced in Chapter 2.

4.4.1.2 Formalism, modelling languages, models, and tools

In the following, based on the MPM ontology we list the formalism, modelling languages, models and tools used for developing CPSs with EBCPS

- Languages and Models
  - Modelling Language: IRM(-SA)Model, DEECoDesignModel (i.e. ensemble part).
  - Java Code: DEECoDesignModel, DEECoRuntimeModel, Self-Adaptation-Model
  - MATSim Code: MATSimDesignModel, MATSimRuntimeModel
  - OMNeT++ Code: OMNeT++DesignModel, OMNeT++RuntimeModel
  - ROS Code: ROSDesignModel, ROSRuntimeModel
- Megamodel
  - Megamodel: DEECoMegaModel (DEECo stands for Dependable Emergent Ensembles of Components)
  - MegamodelFragment: RequirementsMegaModelFragment, DesignMegaModelFragment, RuntimeMegaModelFragment, Self-AdaptationMegaModelFragment, SimulationMegaModelFragment.

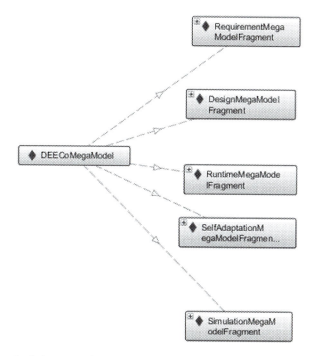

Figure 4.7 An OntoGraf diagram of the megamodel of DEECo as instances of the MPM ontology classes.

⊛ TransformationRelation: Requirements Capturing Operation, Requirements-Design Transformation Operation, Self-Adaptation Capturing Operation, Ensembles Capturing Operation, Simulation Capturing Operation, Instantiation Design-Transformation Operation, Instantiation Simulation-Transformation Operation, Runtime-to-Self-Adaptation Integration Operation, Runtime-to-Vehicle Simulation Integration Operation, Runtime-to-Robot Simulation Integration Operation, Simulation-to-Simulation Integration Operation.

• Modelling Tools
   ⊛ ModelingTool: IRM-SA (Invariant Refinement Method – Self-Adaptation) and EDL (Ensemble Definition Language) developed in D3S Charles University in Prague, EclipseEpsilon
   ⊛ RuntimeTool: jDEECo and its plugins from Department of Distributed and Dependable Systems, Charles University in Prague
   ⊛ SimulationTool: ROS from Open Source Robotics Foundation, OMNet++ from OpenSim Ltd, MATSim from MATSim Community.

4.4.1.3 Simulation stage

As an example, we present the megamodels, transformations and tools for the specific example presented in Chapter 2. The example describes a scenario where an autonomous vehicle joins a road train during its ride from city A to city C. The driver decides to stop in city B in the middle of the trip for sightseeing and the vehicle has to find a parking spot in the city B by reserving in a parking lot or on the spot. The development steps related to this specific example are:

Requirements – MPM Ontology. Developing any system starts with collecting system requirements. We specify requirements using the Invariant Refinement Method (IRM) [117]. For instance, in Fig. 4.8 the goal of the vehicle is to find a parking near to the sightseeing destination in city B. Each vehicle features a role that contains: Vehicle id, the parking lot capacity, the vehicle position, etc. To achieve the goal, the vehicle component must receive the capacity of the nearest parking lot by exchanging data, and execute a process for selecting a suitable parking slot.

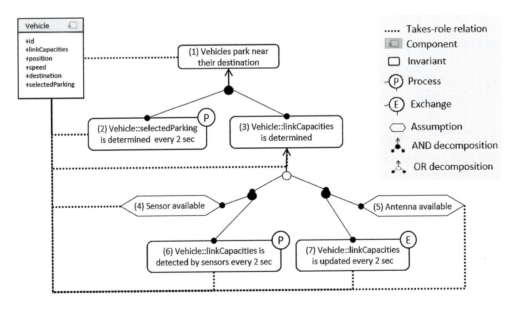

Figure 4.8 IRM tree for a smart parking scenario (based on [118]).

The MegamodelFragment, Models, Transformations and Tools for this example in the requirements step are:

MegaModel Fragment RequirementsMegaModelFragment
- MegamodelFragment: RequirementsMegaModelFragment
  ◦ Model(s): IRMModel

```
@Component
public class Vehicle extends VehicleRole {
    ...
    @Process
    @PeriodicScheduling(period = 2000)
    public static void selectParkingProcess(...) {
        ...
    }
    ...
}


@Ensemble
@PeriodicScheduling(period = 2000)
public class CapacityExchangeEnsemble {

    @Membership
    public static boolean membership(...) {
        ...
        return false;
    }

    @KnowledgeExchange
    public static void map( ... ) {
        ...
    }
}
```

Figure 4.9 Snippet of jDEECo code for the Vehicle Component and the Ensemble between the Vehicle and the Parking lot (based on [118]).

Tools/Models Transformations of RequirementsMegaModelFragment

- Model Transformation: Requirements Capturing Transformation – Modelling requirements refinements by Human
  - Input Model(s): None
  - Output Model(s): IRMModel

Design – MPM Ontology. In this design step, the developer generates the skeleton of system components (i.e. with the roles) and ensembles by transforming an IRM model to jDEECo code [119,120]. The developer has to implement the processes (behaviour) of the vehicle component and parking lot component, and the membership condition and knowledge exchange in the ensembles (Fig. 4.9).

In the example presented in Chapter 2, the plan is to perform a simulation of the vehicle movements on the roads, thus the developer will need to use the plugin for linking MATSim and OMNet++ with the system components and ensembles. The specification of ensembles supports determining communication boundaries over the gossipping-based knowledge dissemination [121].

For more complex representation of the ensembles, the developer can use the Intelligent Ensemble framework[4] [122], where he/she is able to represent hierarchical ensembles with optimisation over forming the ensembles by using EDL. In the example, the intelligent ensemble can be employed in reserving the parking slot by considering the balance in services provided by parking lots [123].

When the autonomous vehicle is on the road, the driver decides to join a road train to save fuel. In this case, the vehicle manages the safe distance between the vehicles in the road train using (Cooperative) Adaptive Cruise Control (C)ACC [124]. To support the self-adaptation in jDEECo, the developer has to add the state machines that represent the modes and the transitions between them [125]. In jDEECo code, the modes are represented as an enumeration and associated to component processes using annotations.

The MegamodelFragments, Models, Transformations and Tools of the example in the design stage are:

MegamodelFragment DesignMegaModelFragment

- MegamodelFragment: DesignMegaModelFragment
    - Model(s): DEECoDesignModel, MATSimDesignModel, OMNeT++DesignModel

Tools/Models Transformations of DesignMegaModelFragment

- Transformation: Requirements-Design Transformation Operation – Capturing requirements refinements in design time
    - Input Model(s): IRMModel
    - Output Model(s): DEECoDesignModel
- Transformation: Self-Adaptation Capturing Operation – Modelling requirements refinements by Human
    - Input Model(s): None
    - Output Model(s): DEECoDesignModel
- Transformation: Ensemble Capturing Operation – Modelling requirements refinements by Human
    - Input Model(s): None
    - Output Model(s): DEECoDesignModel (i.e. ensemble part)
- Transformation: Simulation Capturing Operation – Modelling requirements refinements by Human
    - Input Model(s): None
    - Output Model(s): DEECoDesignModel, MATSimDesignModel, OMNeT++DesignModel

---

[4] http://d3s.mff.cuni.cz/software/deeco/files/seams-2017.zip or http://dx.doi.org/10.4230/DARTS.3.1.6.

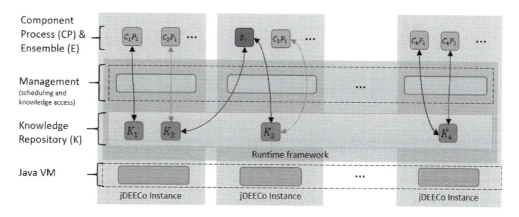

Figure 4.10 jDEECo runtime framework architecture (based on [119]).

Runtime – MPM Ontology. The developer initiates the system components and ensembles to run them in the jDEECo runtime framework [119] (Fig. 4.10). The framework manages scheduling and executing the tasks, and accessing the knowledge repository in which the component's knowledge is stored. More specifically, the framework schedules the processes in the components, forms ensembles and exchanges knowledge between the components. The components and ensembles are distributed and run over different virtual machines, where they access the distributed knowledge repository to execute process and exchange knowledge.

When using intelligent ensembles (Fig. 4.11), the membership condition is evaluated using Z3 SMT solver. The running ensembles are part of the DEECoRuntimeModel.

The MegamodelFragments, Models, Operations and Tools the example in the runtime step are:

MegamodelFragment RuntimeMegaModelFragment

- MegamodelFragment: RuntimeMegaModelFragment
    - Model(s): DEECoRuntimeModel, DEECoRuntime-Self-AdaptationModel, DEECoRuntime-MATSimRuntime-OMNet++RuntimeModel

Tools/Models Transformations of RuntimeMegaModelFragment

- Transformation: Instantiation Design-Transformation Operation – Instantiation of the DEECo components and ensembles
       Input Model(s): DEECoDesignModel
       Output Model(s): DEECoRuntimeModel, Self-AdaptationModel

Self-Adaptation – MPM Ontology. In the design step, we support the self-adaptation in the vehicle to asset the driver in his/her trip. The self-adaptation is represented in jDEECo code using annotations, and the guards are represented as Boolean conditions. At runtime, the evaluation of the guards is performed periodically, and based on the results the component decides if adaptation should be performed.

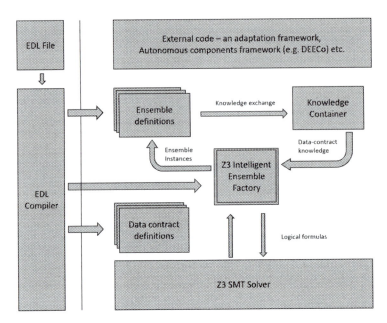

Figure 4.11 Framework high-level architecture that supports ensemble definition with EDL (based on [123]).

MegamodelFragment Self-AdaptationMegaModelFragment

• MegamodelFragment: Self-AdaptationMegaModelFragment

  ◦ Model(s): Self-AdaptationModel

The MegamodelFragment, Models, Transformations and Tools the example in the self-adaptation step are:

Tools/Models Transformations of Self-AdaptationMegaModelFragment

• Transformation: Runtime-to-Self-Adaptation Integration Operation – Integrate the runtime model with the adaptation model

    Input Model(s): DEECoRuntimeModel, Self-AdaptationModel

    Output Model(s): DEECoRuntime-Self-AdaptationModel

Simulation – MPM Ontology. In the design step, the simulation details are provided including the cities' locations, the vehicles and their locations, and the parking lots and their availability. At runtime, the MATSim and OMNet++ are associated to DEECo runtime model. The integration consists of synchronising the models (Fig. 4.12) in order to execute the defined scenario and validates the results, where OMNet++ is able to simulate the delays in the network and MATSim simulate the vehicles.

The MegamodelFragment, Models, Transformations and Tools of the example in the simulation step are:

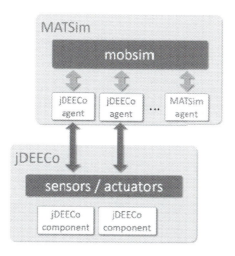

Figure 4.12 jDEECo integration with MATSim.

MegamodelFragment SimulationMegaModelFragment
- MegamodelFragment: SimulationMegaModelFragment
  - ⊛ Model(s): MATSimRuntimeModel, OMNeT++RuntimeModel

Tools/Models Transformations of SimulationMegaModelFragment
- Transformation: Instantiation Simulation-Transformation Operation - Instantiation of the Simulation Models
    Input Model(s): MATSimDesignModel, OMNeT++DesignModel
    Output Model(s): MATSimRuntimeModel, OMNeT++RuntimeModel
- Transformation: Runtime-to-Vehicle Simulation Integration Operation - System Design Validation
    Input Model(s): DEECoRuntimeModel, MATSimRuntimeModel, OM-Net++RuntimeModel
    Output Model(s): DEECoRuntime-MATSimRuntime-OMNet++Runtime-Model

4.4.2 HPI CPSLab
4.4.2.1 Overview

In Fig. 2.23 the tools landscape for developing the aforementioned robots of the CPS production setting of the HPI CPSLab example introduced in Chapter 2 is depicted. It consists of MATLAB®/Simulink for modelling and simulation, dSPACE SystemDesk for modelling software architecture, hardware configuration, and task mapping, dSPACE TargetLink for code generation and the FESTO Robotino-Library with the FESTO Robotino©Sim simulator.

The HPI CPSLab development methodology employs a number of megamodel fragments used at each stage of its methodology. Fig. 4.13 depicts the various megamodel fragments.

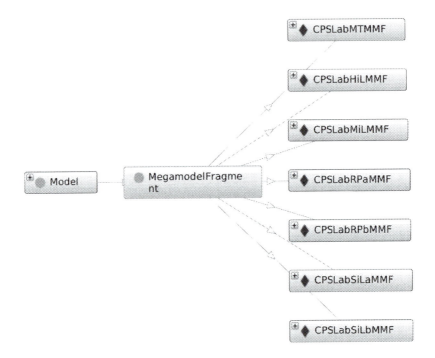

Figure 4.13 All megamodel fragments of the CPSLabMM megamodel.

An overview of the stages and activities with an emphasis on models, tools, and multi-paradigm modelling is depicted in Fig. 4.14. The figure graphically shows the toolchain and languages used at each stage of the development process and their dependencies.

In the following, we outline which elements of the megamodel refer to which MPM ontology elements and how the megamodel fragments cover the different scenarios associated with the development process.

4.4.2.2 Formalism, languages, models, and tools

- Languages and Models
    - MATLAB/Simulink Language: ControlModel, PlantModel
    - FESTO Robotino©Sim Language: RobotModel
    - AUTOSAR Language: SystemModel

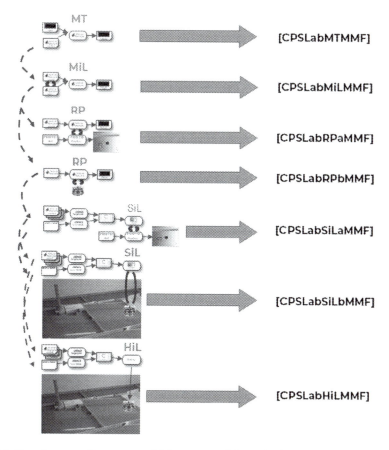

Figure 4.14 Overview over the megamodel fragments of the CPSLab megamodel and how the models are related (dashed arrows).

- Megamodel
  - Megamodel: CPSLabMM
  - MegamodelFragments: CPSLabMTMMF, CPSLabMiLMMF, CPSLabRPaMMF, CPSLabRPbMMF, CPSLabSiLaMMF, CPSLabSiLbMMF, CPSLabHiLMMF
  - ModelRelations: (see detailed definition of the mega-model fragments)
- Tool
  - SimulationTool: MATLAB Stateflow Simulator
  - TransformationTool: dSPACE TargetLink
  - ModelingTool: dSPACE SystemDesk
  - SimulationTool: FESTO Robotino©Sim
  - VisualizationTool: FESTO Robotino©View
  - ExecutionTool: Execution on a Desktop computer

- ExecutionTool: Remote execution on a Robotino©Robot
- ExecutionTool: Local execution on a Robotino Robot

4.4.3 Simulation stage

The first step of the methodology of Fig. 2.19 is a simulation stage that focuses on the model development resp. functional development for the employed control laws. At this stage, many details resulting from the physical and cyber parts of the system are ignored resp. simplified such as real sensor values with noise, specific effects of scheduling, the impact of communication interaction and messages, and timing/memory/computation constraints.

In the simulation stage we have two development activities: model test (MT) and model-in-the-loop (MiL). We will now outline how they can be supported with megamodel fragments consisting of model instances and tool applications.

4.4.3.1 Model Test (MT)

The model test activity is depicted in Fig. 4.15. It is rather trivial as it only employs a single model of the control algorithm plus some auxiliary models for test inputs and expected test results. As depicted in Fig. 4.16, the model of the control algorithm is simulated by employing the test inputs and comparing the simulation results with the expected results.

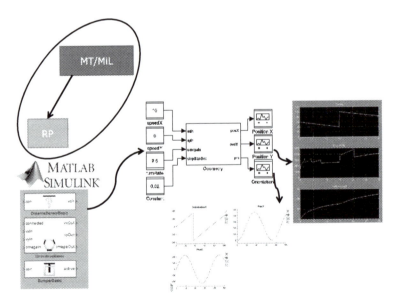

Figure 4.15 Overview of the model test in the simulation stage of [126].

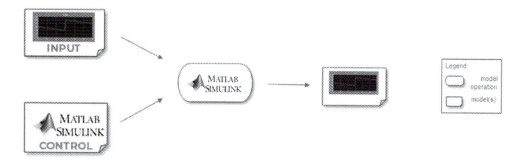

Figure 4.16 Model Test activity.

4.4.3.2 Model Test – MPM ontology

MegamodelFragment CPSLabMTMMF
- MegamodelFragment: CPSLabMTMMF
  - ∘ Model(s): ControlModel

Tools/Models Transformations of CPSLabMiLMMF
- Transformation: One Shot Simulation
  - ∘ Input Model(s): ControlModel, Input data (entered with MATLAB/Stateflow Simulator)
  - ∘ Output Model(s): Output data (visualised with MATLAB/Stateflow Simulator)
  - ∘ Modelling Tool: MATLAB/Stateflow Simulator

In Fig. 4.17, the added Megamodel Fragment CPSLabMTMMF and its elements for the CPSLab ontology outlined in the text are presented.

4.4.3.3 Model-in-the-loop

In a second step, the model of the control behaviour is combined with a MAT-LAB/Simulink model of the plant by means of a *model-in-the-loop* (MiL) simulation as shown in Fig. 4.18, which uses the feedback provided by the plant model to evaluate that the control behaviour is as expected.

In contrast to model test, model-in-the-loop simulation employs besides a model of the control algorithm a model of the plant and as depicted in Fig. 4.19 uses simulation to explore how well both models fit together.

4.4.3.4 Model-in-the-loop – MPM ontology

MegamodelFragment CPSLabMiLMMF
- MegamodelFragment: CPSLabMiLMMF
  - ∘ Model(s): ControlModel, PlantModel

Tools/Models Transformations of CPSLabMiLMMF
- Transformation: Model-in-the-Loop Simulation

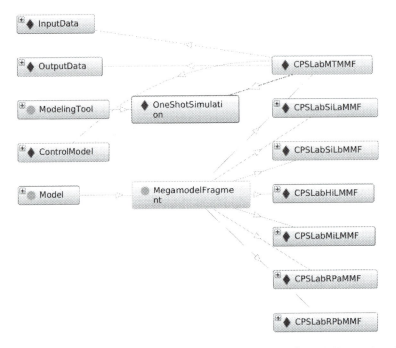

Figure 4.17 Part of the ontology for the megamodel fragment CPSLabMTMMF covering the Model Test activity.

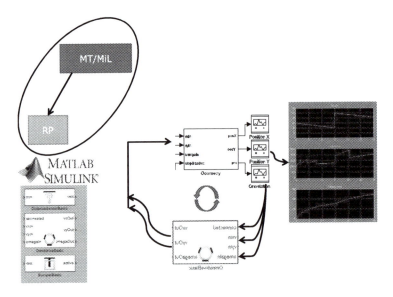

Figure 4.18 Overview of the model in loop simulation activity of the simulation stage.

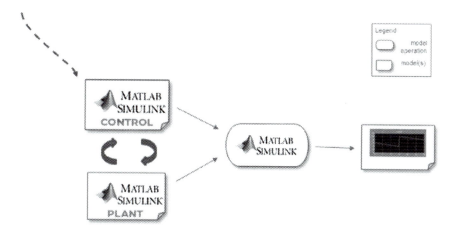

Figure 4.19 Model-in-the-Loop activity.

- Input Model(s): ControlModel, PlantModel
- Output Model(s): Output data (visualised with MATLAB/Stateflow Simulator)
- Modelling Tool: MATLAB/Stateflow Simulator

Fig. 4.20 depicts the added megamodel fragment CPSLabMiLMMF and its elements for the CPSLab ontology.

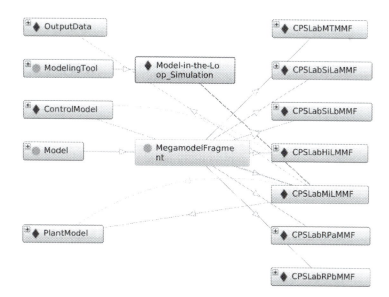

Figure 4.20 Part of the ontology for the megamodel fragment CPSLabMiLMMF covering the Model-in-the-Loop activity.

4.4.3.5 Rapid prototyping

As the validity of plant models is often rather limited when it comes to sophisticated aspects of the physical behaviour, as an additional step *Rapid Prototyping* (RP) as depicted in Fig. 4.21 is supported.

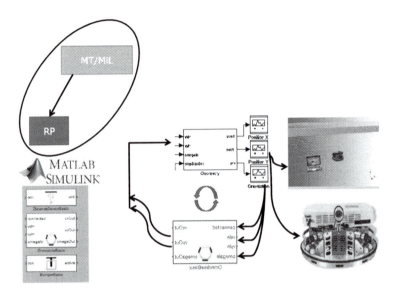

Figure 4.21 Overview of the rapid prototyping activity of the simulation stage.

Rapid prototyping is supported in two forms. In the first case, it is performed by employing a sophisticated simulator for the robot as depicted in Fig. 4.22. While not necessarily exposing the control algorithm to physical reality as far as captured by the sophisticated model of the robot, the simulator already captures much more details than the plant model while still allowing one to analyse the behaviour much more easily than in the case of using a real robot.

For the second case, which is applicable to smaller control behaviour, the model of the control behaviour is connected to the real robot as depicted in Fig. 4.23. In this case, real sensor values with noise and timing constraints of the environment and platform are taken into account. However, specific effects of scheduling, the impact of communication interaction and messages, and memory/computation constraints remain uncovered. In addition, performing analyses in this configuration might be difficult since running the algorithm against the robot is more difficult than running it against a simulator.

For larger scenarios including several robots, a link to a real hardware setup is not feasible. Instead we employ a model-in-the-loop (MiL) simulation where a complex environment and the communication between the robots can be explored. While this

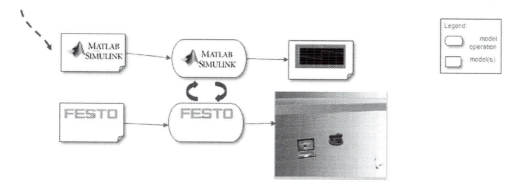

Figure 4.22 Rapid Prototyping using a detailed robot simulator.

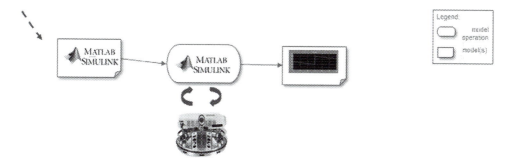

Figure 4.23 Rapid Prototyping with a remotely controlled robot.

covers the impact of communication interaction and messages, other aspects like real sensor values with noise, specific effects of scheduling and timing/memory/computation constraints are, however, not covered.

4.4.3.6 Rapid prototyping – MPM ontology

In Fig. 4.24, the added megamodel fragment CPSLabRPaMMF and its elements for the CPSLab ontology outlined in the text are presented.

MegamodelFragment CPSLabRPaMMF
- MegamodelFragment: CPSLabRPaMMF
  - Model(s): ControlModel, RobotModel

Tools/Models Transformations of CPSLabRPaMMF
- Transformation: Rapid Prototyping with Robot Simulation
  - Input Model(s): ControlModel, RobotModel
  - Output Model(s): Output data (visualised with MATLAB/Stateflow Simulator and/or FESTO Robotino©View)
  - Modelling Tool: MATLAB/Stateflow Simulator, FESTO Robotino©Sim

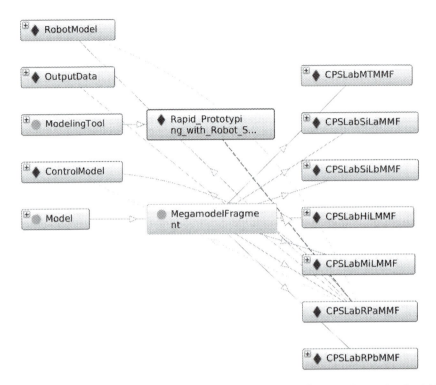

Figure 4.24 Part of the ontology for the megamodel fragment CPSLabRPaMMF covering Rapid Prototyping with robot simulation.

MegamodelFragment CPSLabRPbMMF
- MegamodelFragment: CPSLabRPbMMF
  - Model(s): ControlModel

Tools/Models Transformations of CPSLabRPbMMF
- Transformation: Rapid Prototyping with Robot Execution
  - Input Model(s): ControlModel
  - Output Model(s): Output data (visualised with MATLAB/Stateflow Simulator and/or observed)
  - Modelling Tool: MATLAB/Stateflow Simulator, Remote execution on a Robotino©Robot

The added megamodel fragment CPSLabRPbMMF and its elements for the CPSLab ontology outlined in the text are depicted also in Fig. 4.25.

4.4.3.7 Prototyping stage

The second supported stage is the prototyping stage where the focus changes from models to their implementation in software or hardware and where besides the individual

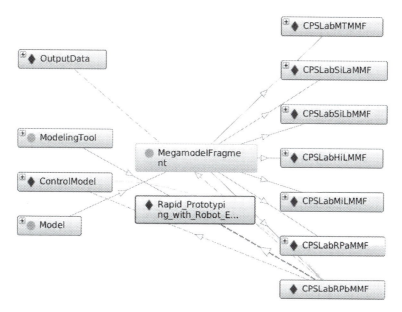

Figure 4.25 Part of the ontology for megamodel fragment CPSLabRPbMMF covering Rapid Prototyping with Robot Execution.

functions the system architecture is also covered. Due to this refined view, in particular discretisation effects of the cyber part that are absent in the abstract mathematical models employed in the former stage now become visible.

At this stage, less details are ignored resp. simplified as step by step specific effects of scheduling, the impact of communication interaction and messages, and timing/memory/computation constraints are taken into account.

During this stage, the models must be refined such that besides the individual functions, the system architecture is also defined. As depicted in Fig. 4.26, this is done by first defining components and their communication via port types, messages, interfaces, and data types with AUTOSAR and by mapping the beforehand considered functional parts on them. In this step, we also have to map the functionality extending the existing models and where necessary add custom implementation files.

In a second step, we then define the overall architecture using AUTOSAR including task specification and the hardware configuration besides the components and their communication. As depicted in Fig. 4.27, an important element of this refinement is real-time constraints, e.g. to guarantee safety constraints. A combination of hard and soft real-time aspects at the functional as well as the architectural levels must be defined including a mapping to hard and soft real-time task with proper levels for the priorities.

Concerning the verification, we employ code generation at the prototyping stage and try to step by step add more and more details of the software and hardware to the picture in the following steps.

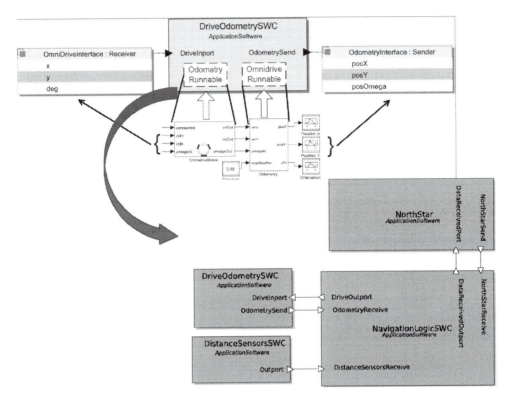

Figure 4.26 Overview of the definition of the software architecture for the prototyping stage (from [127]).

4.4.3.8 Software in the Loop (SiL)

The *Software-in-the-Loop* (SiL) simulation at the prototyping stage as depicted in Fig. 4.28 requires that code generation is employed to derive code for the functional models and architectural models. In special cases, also additional manually developed code has to be integrated. Then the code is executed and run against the available simulation of the robot and its environment.

As we still do not always use the real hardware, we still ignore resp. simplify elements such as real sensor values with noise and timing constraints of the environment or platform not covered by the simulator. However, specific effects of scheduling, the impact of communication interaction and messages, and by the simulator covered timing constraints of the environment or platform, and timing/memory/computation constraints of the software.

SiL can actually be performed in different ways: A first version executes the generated software on a desktop computer and runs it against a simulator as depicted in Fig. 4.29.

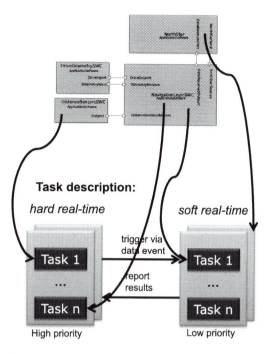

Figure 4.27 Overview of the mapping of the architecture to tasks and communication for the prototyping stage.

A second form executes the generated software on a desktop computer and in contrast to the first form links it to the robot as depicted in Fig. 4.30.

4.4.3.9 Software in the Loop (SiL) – MPM ontology

MegamodelFragment CPSLabSiLaMMF
- MegamodelFragment: CPSLabSiLaMMF
   - Model(s): ControlModel★, SystemModel, RobotModel

Tools/Models Transformations of CPSLabSiLaMMF
- Transformation: FunctionCodeGeneration★
   - Input Model(s): ControlModel
   - Output Model(s): ControlCode
   - Modelling Tool: dSPACE SystemDesk
- Transformation: SystemCodeGeneration
   - Input Model(s): SystemModel
   - Output Model(s): SystemCode
   - Modelling Tool: dSPACE SystemDesk

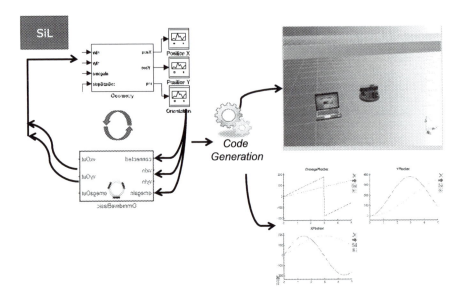

Figure 4.28 Overview of software-in-the-loop simulation of the prototyping stage.

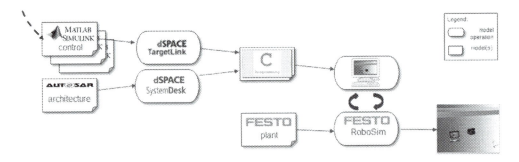

Figure 4.29 Software in the Loop with desktop computer and simulator.

- Transformation: Software-in-the-Loop Simulation
  - Input Model(s): ControlCode★, SystemCode, RobotModel
  - Output Model(s): Output data (visualised with MATLAB/Stateflow Simulator and/or FESTO Robotino©View)
  - Modelling Tool: Execution on a Desktop computer, FESTO Robotino©Sim

The added megamodel fragment CPSLabSiLaMMF and its elements for the CPSLab ontology outlined in the text are depicted also in Fig. 4.31.

MegamodelFragment CPSLabSiLbMMF

- MegamodelFragment: CPSLabSiLbMMF
  - Model(s): ControlModel★, SystemModel

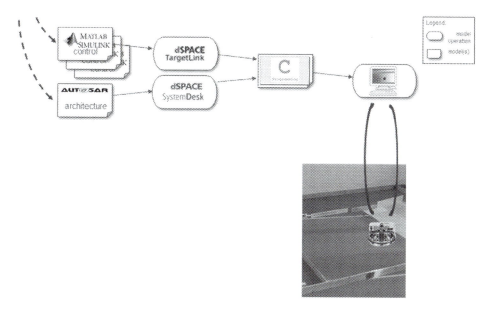

Figure 4.30 Software in the Loop with desktop computer and robot.

Tools/Models Transformations of CPSLabSiLbMMF
- Transformation: FunctionCodeGeneration⋆
  - Input Model(s): ControlModel
  - Output Model(s): ControlCode
  - Modelling Tool: dSPACE TargetLink
- Transformation: SystemCodeGeneration
  - Input Model(s): SystemModel
  - Output Model(s): SystemCode
  - Modelling Tool: dSPACE SystemDesk
- Transformation: Software-in-the-Loop Execution
  - Input Model(s): ControlCode⋆, SystemCode
  - Output Model(s): Output data (visualised with MATLAB/Stateflow Simulator and/or observed)
  - Modelling Tool: Remote execution on a Robotino©Robot, Execution on a Desktop computer

The added megamodel fragment CPSLabSiLbMMF and its elements for the CPSLab ontology outlined in the text are depicted graphically in Fig. 4.32.

4.4.3.10 Hardware in the Loop (HiL)

By moving on to the lab itself, we can then also consider a *hardware-in-the-loop* (HiL) simulation at the prototyping stage as sketched in Fig. 4.33. Besides the software that

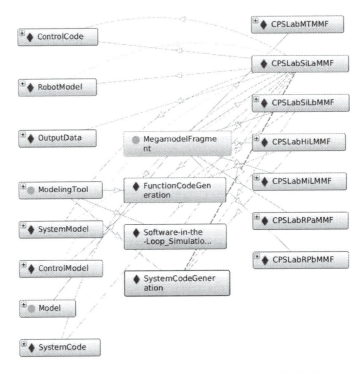

Figure 4.31 Part of the ontology for the megamodel fragment CPSLabSiLaMMF covering SiL with Simulation.

is generated or integrated, the specific characteristics of the robot hardware and lab environment and its hardware can also be experienced.

As we now employ the real hardware, we no longer ignore resp. simplify any elements. Therefore, now real sensor values with noise, specific effects of scheduling, the impact of communication interaction and messages, and timing/memory/computation constraints are all considered.

HiL links the generated software such that it can be executed on the robot as depicted in Fig. 4.34.

4.4.3.11 Hardware in the Loop (HiL) – MPM ontology

MegamodelFragment CPSLabHiLMMF
- MegamodelFragment: CPSLabHiLMMF
  - Model(s): ControlModel★, SystemModel
Tools/Models Transformations of CPSLabHiLMMF
- Transformation: FunctionCodeGeneration★
  - Input Model(s): ControlModel
  - Output Model(s): ControlCode

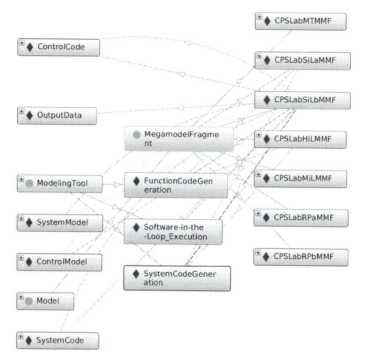

Figure 4.32 Part of the ontology for the megamodel fragment CPSLabSiLbMMF covering SiL with Execution.

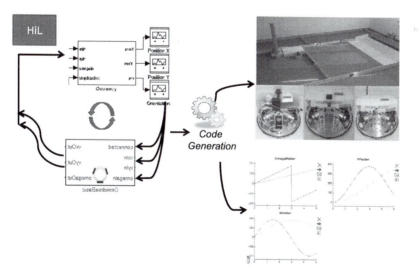

Figure 4.33 Overview of hardware in the loop (HiL) testing in the prototyping stage of [126].

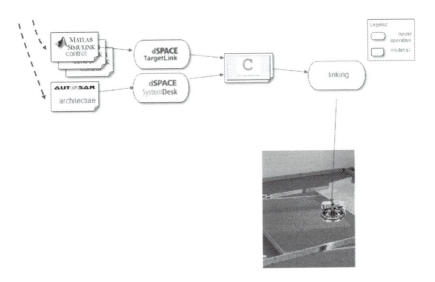

Figure 4.34 Hardware in the Loop (HiL).

  * Modelling Tool: dSPACE TargetLink
* Transformation: SystemCodeGeneration
  * Input Model(s): SystemModel
  * Output Model(s): SystemCode
  * Modelling Tool: dSPACE SystemDesk
* Transformation: Hardware-in-the-Loop Execution
  * Input Model(s): ControlCode*, SystemCode
  * Output Model(s): Output data (observed)
  * Modelling Tool: Local execution on a Robotino©Robot

The added megamodel fragment CPSLabHiLMMF and its elements for the CPSLab ontology outlined in the text are depicted graphically in Fig. 4.35.

4.5. Conclusion

In this chapter we first presented a thorough state of the art treatment on the notions required for MPM such as core modelling notions as well as multi-formalism and global model management approaches, languages and tools. All these elements are an essential component to support MPM. We characterised model management approaches according to their modularity and incremental execution properties as required to scale for the large complex CPS we face today. We studied these modularity and incremental execution properties for the three main axis of model and modelling languages, model operations / transformations and model management. It shows that despite that many

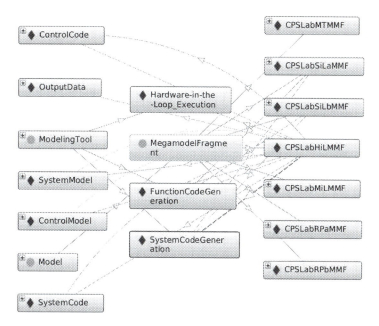

Figure 4.35 Part of the ontology for the megamodel fragment CPSLabHiLMMF covering the Hardware-in-the-Loop activity.

approaches exist, most of them only address part of the problem therefore pointing to the need for a global model management framework.

We then presented an overview of the MPM ontology developed during the MPM4CPS COST action, whose complete specification and artifacts are available from the MPM4CPS website [114]. We introduced the main classes and properties of the ontology, whose specification is grounded on the state of the art to produce a first version of the ontology. The ontology was then validated and refined by modelling the MPM aspects of the EBCPS and HPI CPSLab development environments. This shows the capability of our MPM ontology to model these environments.

Our ontology is expressed using the Web Ontology Language (OWL), following the MPM principle that advocates using the most appropriate language for the modelling activity to be performed. As presented in Chapter 2, given the exploratory modelling nature of this work, OWL, which is one of the best languages for developing classifications was chosen. Its richness and precise semantics based on set theory provides extensive reasoning capabilities that we plan to investigate more to improve our classification of the MPM domain, and for the next constructive phase following this work that will consist of developing a solution for implementing model management for MPM based on the knowledge captured by the ontology.

References

[1] P.J. Mosterman, H. Vangheluwe, Computer automated multi-paradigm modeling: an introduction, Simulation 80 (9) (2004) 433–450.

[2] Jean Bézivin, Frédéric Jouault, Peter Rosenthal, Patrick Valduriez, Modeling in the large and modeling in the small, in: Model Driven Architecture, in: Lecture Notes in Computer Science (LNCS), vol. 3599/2005, Springer-Verlag, 2005, pp. 33–46.

[3] Jean-Marie Favre, Foundations of model (driven) (reverse) engineering – episode I: story of the fidus papyrus and the solarus, in: Post-Proceedings of Dagstuhl Seminar on Model Driven Reverse Engineering, 2004.

[4] D. Harel, B. Rumpe, Modeling languages: syntax, semantics and all that stuff, part I: the basic stuff, Technical Report, ISR, 2000.

[5] Holger Giese, Tihamer Levendovszky, Hans Vangheluwe (Eds.), Summary of the Workshop on Multi-Modelling Paradigms: Concepts and Tools, 2006.

[6] David Broman, Edward A. Lee, Stavros Tripakis, Martin Törngren, Viewpoints, formalisms, languages, and tools for cyber-physical systems, in: Proceedings of the 6th International Workshop on Multi-Paradigm Modeling, MPM '12, ACM, New York, NY, USA, 2012, pp. 49–54.

[7] ISO/IEC/IEEE 42010:2011. Systems and software engineering - architecture description, the latest edition of the original IEEE std 1471:2000, recommended practice for architectural description of software-intensive systems, 2011.

[8] Fatma Dandashi, Vinay Lakshminarayan, Nancy Schult, Multiformalism, multiresolution, multiscale modeling, in: 2015 Winter Simulation Conference (WSC), 2015, pp. 2622–2631.

[9] Hassan Reza, Emanuel S. Grant, Model oriented software architecture, in: Proceedings of the 28th Annual International Computer Software and Applications Conference (COMPSAC), 2004, pp. 4–5.

[10] Mauro Iacono, Marco Gribaudo, An introduction to multiformalism modeling, in: Marco Gribaudo, Mauro Iacono (Eds.), Theory and Application of Multi-Formalism Modeling, IGI Global, Hershey, 2014, pp. 1–16.

[11] S. Lacoste-Julien, H. Vangheluwe, J. De Lara, P.J. Mosterman, Meta-modelling hybrid formalisms, 2004, pp. 65–70.

[12] H. Vangheluwe, J. De Lara, Computer automated multi-paradigm modelling: meta-modelling and graph transformation, vol. 1, 2003, pp. 595–603.

[13] B.P. Zeigler, H. Praehofer, Interfacing continuous and discrete models for simulation and control, SAE Technical Papers, 1998.

[14] B.P. Zeigler, Embedding DEV&DESS in DEVS: characteristic behaviors of hybrid models, 2006, pp. 125–132.

[15] Enrico Barbierato, Marco Gribaudo, Mauro Iacono, Modeling hybrid systems in {SIMTHESys}, Electronic Notes in Theoretical Computer Science 327 (2016) 5–25.

[16] Enrico Barbierato, Marco Gribaudo, Mauro Iacono, Simulating Hybrid Systems Within SIMTHESys Multi-formalism Models, Springer International Publishing, Cham, 2016, pp. 189–203.

[17] S. Balsamo, G. Dei Rossi, A. Marin, A survey on multi-formalism performance evaluation tools, 2012, pp. 15–23.

[18] Kishor S. Trivedi, SHARPE 2002: symbolic hierarchical automated reliability and performance evaluator, in: DSN '02: Proceedings of the 2002 International Conference on Dependable Systems and Networks, IEEE Computer Society, Washington, DC, USA, 2002, p. 544.

[19] G. Ciardo, A.S. Miner, SMART: the stochastic model checking analyzer for reliability and timing, in: Quantitative Evaluation of Systems, 2004. QEST 2004. Proceedings. First International Conference on the, Sept 2004, pp. 338–339.

[20] Gianfranco Ciardo, Andrew S. Miner, Min Wan, Advanced features in SMART: the stochastic model checking analyzer for reliability and timing, SIGMETRICS Performance Evaluation Review 36 (4) (March 2009) 58–63.

[21] G. Ciardo, R.L. Jones III, A.S. Miner, R.I. Siminiceanu, Logic and stochastic modeling with SMART, Performance Evaluation 63 (June 2006) 578–608.

[22] Falko Bause, Peter Buchholz, Peter Kemper, A toolbox for functional and quantitative analysis of DEDS, in: Proceedings of the 10th International Conference on Computer Performance Evaluation: Modelling Techniques and Tools, TOOLS '98, Springer-Verlag, London, UK, 1998, pp. 356–359.

[23] W.H. Sanders, Integrated frameworks for multi-level and multi-formalism modeling, in: Petri Nets and Performance Models, 1999. Proceedings. The 8th International Workshop on, 1999, pp. 2–9.

[24] Graham Clark, Tod Courtney, David Daly, Dan Deavours, Salem Derisavi, Jay M. Doyle, William H. Sanders, Patrick Webster, The Mobius modeling tool, in: Proceedings of the 9th International Workshop on Petri Nets and Performance Models (PNPM'01), IEEE Computer Society, Washington, DC, USA, 2001, p. 241.

[25] Tod Courtney, Shravan Gaonkar, Ken Keefe, Eric Rozier, William H. Sanders, Möbius 2.3: an extensible tool for dependability, security, and performance evaluation of large and complex system models, in: DSN, IEEE, 2009, pp. 353–358.

[26] Daniel D. Deavours, Graham Clark, Tod Courtney, David Daly, Salem Derisavi, Jay M. Doyle, William H. Sanders, Patrick G. Webster, The Möbius framework and its implementation, 2002.

[27] Juan de Lara, Hans Vangheluwe, AToM3: a tool for multi-formalism and meta-modelling, in: Ralf-Detlef Kutsche, Herbert Weber (Eds.), FASE, in: Lecture Notes in Computer Science, vol. 2306, Springer, 2002, pp. 174–188.

[28] M. Del, V. Sosa, S.T. Acuna, J. De Lara, Metamodeling and multiformalism approach applied to software process using AToM [Enfoque de metamodelado y multiformalismo aplicado al proceso software usando AToM3], 2007, pp. 367–374.

[29] F. Franceschinis, M. Gribaudo, M. Iacono, N. Mazzocca, V. Vittorini, Towards an object based multi-formalism multi-solution modeling approach, in: Daniel Moldt (Ed.), Proc. of the Second International Workshop on Modelling of Objects, Components, and Agents (MOCA'02), Aarhus, Denmark, August 26–27, 2002, aug 2002, pp. 47–66, Technical Report DAIMI PB-561.

[30] V. Vittorini, G. Franceschinis, M. Gribaudo, M. Iacono, N. Mazzocca, DrawNet++: model objects to support performance analysis and simulation of complex systems, in: Proc. of the 12th Int. Conference on Modelling Tools and Techniques for Computer and Communication System Performance Evaluation (TOOLS 2002), London, UK, April 2002.

[31] Giuliana Franceschinis, Marco Gribaudo, Mauro Iacono, Stefano Marrone, Nicola Mazzocca, Valeria Vittorini, Compositional modeling of complex systems: contact center scenarios in OsMoSys, in: ICATPN'04, 2004, pp. 177–196.

[32] G. Franceschinis, M. Gribaudo, M. Iacono, S. Marrone, F. Moscato, V. Vittorini, Interfaces and binding in component based development of formal models, in: Proceedings of the Fourth International ICST Conference on Performance Evaluation Methodologies and Tools, VALUETOOLS '09, ICST, Brussels, Belgium, 2009, pp. 44:1–44:10, ICST (Institute for Computer Sciences, Social-Informatics and Telecommunications Engineering).

[33] G. Gribaudo, M. Iacono, M. Mazzocca, V. Vittorini, The OsMoSys/DrawNET Xe! languages system: a novel infrastructure for multi-formalism object-oriented modelling, in: ESS 2003: 15th European Simulation Symposium and Exhibition, 2003.

[34] Enrico Barbierato, Marco Gribaudo, Mauro Iacono, Defining formalisms for performance evaluation with SIMTHESys, Electronic Notes in Theoretical Computer Science 275 (2011) 37–51.

[35] Enrico Barbierato, Marco Gribaudo, Mauro Iacono, A performance modeling language for big data architectures, in: Webjorn Rekdalsbakken, Robin T. Bye, Houxiang Zhang (Eds.), ECMS, European Council for Modeling and Simulation, 2013, pp. 511–517.

[36] Mauro Iacono, Marco Gribaudo, Element based semantics in multi formalism performance models, in: MASCOTS, 2010, pp. 413–416.

[37] M. Iacono, E. Barbierato, M. Gribaudo, The SIMTHESys multiformalism modeling framework, Computers and Mathematics with Applications 64 (2012) 3828–3839.

[38] Mauro Pezze, Michal Young, Constructing multi-formalism state-space analysis tools: using rules to specify dynamic semantics of models, 1997, pp. 239–249.

[39] Enrico Barbierato, Gian-Luca Dei Rossi, Marco Gribaudo, Mauro Iacono, Andrea Marin, Exploiting product forms solution techniques in multiformalism modeling, Electronic Notes in Theoretical Computer Science 296 (2013) 61–77.

[40] C.-V. Bobeanu, E.J.H. Kerckhoffs, H. Van Landeghem, Modeling of discrete event systems: a holistic and incremental approach using Petri nets, ACM Transactions on Modeling and Computer Simulation 14 (4) (2004) 389–423.

[41] J.T. Bradley, M.C. Guenther, R.A. Hayden, A. Stefanek, GPA: a multiformalism, multisolution approach to efficient analysis of Large-Scale population models, 2013.

[42] Aniello Castiglione, Marco Gribaudo, Mauro Iacono, Francesco Palmieri, Exploiting mean field analysis to model performances of big data architectures, Future Generations Computer Systems 37 (2014) 203–211.

[43] A.H. Levis, B. Yousefi, Multi-formalism modeling for evaluating the effect of cyber exploits, 2014, pp. 541–547.

[44] A.M. Abusharekh, A.H. Levis, Performance evaluation of SOA in clouds, 2016, pp. 614–620.

[45] Mauro Iacono, Stefano Marrone, Model-based availability evaluation of composed web services, Journal of Telecommunications and Information Technology 4 (2014) 5–13.

[46] A.I. Hernandez, V. Le Rolle, A. Defontaine, G. Carrault, A multiformalism and multiresolution modelling environment: application to the cardiovascular system and its regulation, Philosophical Transactions of the Royal Society A: Mathematical, Physical and Engineering Sciences 367 (1908) (2009) 4923–4940.

[47] S. Chiaradonna, P. Lollini, F.D. Giandomenico, On a modeling framework for the analysis of inter-dependencies in electric power systems, 2007, pp. 185–194.

[48] Francesco Flammini, Stefano Marrone, Mauro Iacono, Nicola Mazzocca, Valeria Vittorini, A multiformalism modular approach to ERTMS/ETCS failure modelling, International Journal of Reliability, Quality and Safety Engineering 21 (01) (2014) 1450001 (pp. 1–29).

[49] Marco Gribaudo, Mauro Iacono, Stefano Marrone, Exploiting Bayesian networks for the analysis of combined attack trees, in: Proceedings of the Seventh International Workshop on the Practical Application of Stochastic Modelling (PASM), Electronic Notes in Theoretical Computer Science 310 (2015) 91–111.

[50] Enrico Barbierato, Marco Gribaudo, Mauro Iacono, Stefano Marrone, Performability modeling of exceptions-aware systems in multiformalism tools, in: ASMTA, 2011, pp. 257–272.

[51] Enrico Barbierato, Andrea Bobbio, Marco Gribaudo, Mauro Iacono, Multiformalism to support software rejuvenation modeling, in: ISSRE Workshops, IEEE, 2012, pp. 271–276.

[52] Enrico Barbierato, Marco Gribaudo, Mauro Iacono, Performance evaluation of NoSQL big-data applications using multi-formalism models, Future Generations Computer Systems 37 (2014) 345–353.

[53] Daniele Codetta Raiteri, Mauro Iacono, Giuliana Franceschinis, Valeria Vittorini, Repairable fault tree for the automatic evaluation of repair policies, in: DSN, 2004, pp. 659–668.

[54] Enrico Barbierato, Marco Gribaudo, Mauro Iacono, Exploiting multiformalism models for testing and performance evaluation in SIMTHESys, in: Proceedings of 5th International ICST Conference on Performance Evaluation Methodologies and Tools - VALUETOOLS 2011, 2011.

[55] A. Qamar, S.J.I. Herzig, C.J.J. Paredis, M. Torngren, Analyzing semantic relationships between multiformalism models for inconsistency management, 2015, pp. 84–89.

[56] Jean Bézivin, Frédéric Jouault, Patrick Valduriez, On the need for megamodels, in: Proceedings of the OOPSLA/GPCE: Best Practices for Model-Driven Software Development workshop, 19th Annual ACM Conference on Object-Oriented Programming, Systems, Languages, and Applications, 2004.

[57] CoEST Project Homepage, http://www.coest.org/. (Accessed 2016).

[58] AMW Project Homepage, https://projects.eclipse.org/projects/modeling.gmt.amw/, 2015.

[59] Epsilon Project Homepage, http://eclipse.org/epsilon/, 2020.

[60] Jean-Marie Favre, Ralf Lämmel, Andrei Varanovich, Modeling the linguistic architecture of software products, in: Proceedings of the 15th International Conference on Model Driven Engineering Languages and Systems, MODELS'12, Springer-Verlag, Berlin, Heidelberg, 2012, pp. 151–167.

[61] Henrik Lochmann, Anders Hessellund, An integrated view on modeling with multiple domain-specific languages, in: Proceedings of the IASTED International Conference Software Engineering (SE 2009), ACTA Press, February 2009, pp. 1–10.

[62] Sebastian J.I. Herzig, Ahsan Qamar, Christiaan J.J. Paredis, An approach to identifying inconsistencies in model-based systems engineering, in: 2014 Conference on Systems Engineering Research, Procedia Computer Science 28 (2014) 354–362.

[63] Gabor Simko, Tihamer Levendovszky, Sandeep Neema, Ethan Jackson, Ted Bapty, Joseph Porter, Janos Sztipanovits, Foundation for model integration: semantic backplane, in: ASME 2012 International Design Engineering Technical Conferences and Computers and Information in Engineering Conference, American Society of Mechanical Engineers, 2012, pp. 1077–1086.

[64] Mark Boddy, Martin Michalowski, August Schwerdfeger, Hazel Shackleton, Steve Vestal, Adventium Enterprises, FUSED: a tool integration framework for collaborative system engineering, in: Analytic Virtual Integration of Cyber-Physical Systems Workshop, 2011.

[65] MoTE Project Homepage, http://www.mdelab.org/mdelab-projects/mote-a-tgg-based-model-transformation-engine/, 2015.

[66] Andreas Seibel, Stefan Neumann, Holger Giese, Dynamic hierarchical mega models: comprehensive traceability and its efficient maintenance, Software and Systems Modeling 9 (4) (2010) 493–528.

[67] Andreas Seibel, Regina Hebig, Holger Giese, Traceability in model-driven engineering: efficient and scalable traceability maintenance, in: Jane Cleland-Huang, Orlena Gotel, Andrea Zisman (Eds.), Software and Systems Traceability, Springer, London, 2012, pp. 215–240.

[68] Thomas Beyhl, Regina Hebig, Holger Giese, A model management framework for maintaining traceability links, in: Stefan Wagner, Horst Lichter (Eds.), Software Engineering 2013 Workshopband, Aachen, in: Lecture Notes in Informatics (LNI), vol. P-215, February 2013, pp. 453–457, Gesellschaft für Informatik (GI).

[69] D. Langsweirdt, N. Boucke, Yolande Berbers, Architecture-driven development of embedded systems with ACOL, in: Object/Component/Service-Oriented Real-Time Distributed Computing Workshops (ISORCW), 2010 13th IEEE International Symposium on, May 2010, pp. 138–144.

[70] Anders Hessellund, Andrzej Wasowski, Interfaces and metainterfaces for models and metamodels, in: Krzysztof Czarnecki, Ileana Ober, Jean-Michel Bruel, Axel Uhl, Markus Wolter (Eds.), Model Driven Engineering Languages and Systems, in: Lecture Notes in Computer Science, vol. 5301, Springer, Berlin Heidelberg, 2008, pp. 401–415.

[71] Stefan Jurack, Gabriele Taentzer, A component concept for typed graphs with inheritance and containment structures, in: Graph Transformations - 5th International Conference, ICGT 2010. Proceedings, Enschede, The Netherlands, September 27–October 2, 2010, 2010, pp. 187–202.

[72] SmartEMF Project Homepage, http://www.itu.dk/~hessellund/smartemf/, 2008.

[73] Composite EMF Models Project Homepage, http://www.mathematik.uni-marburg.de/~swt/compoemf/. (Accessed 2015).

[74] MontiCore Project Homepage, http://www.monticore.de/, 2008.

[75] Arvid Butting, Robert Eikermann, Oliver Kautz, Bernhard Rumpe, Andreas Wortmann, Systematic composition of independent language features, Journal of Systems and Software 152 (2019) 50–69.

[76] Arvid Butting, Robert Eikermann, Oliver Kautz, Bernhard Rumpe, Andreas Wortmann, Modeling language variability with reusable language components, in: SPLC '18, Proceedings of the 22nd International Systems and Software Product Line Conference - Volume 1, Association for Computing Machinery, New York, NY, USA, 2018, pp. 65–75.

[77] Kompren Project Homepage, http://people.irisa.fr/Arnaud.Blouin/software_kompren.html, 2014.

[78] Kompose Project Homepage, http://www.kermeta.org/mdk/kompose, 2009.

[79] Anne Etien, Alexis Muller, Thomas Legrand, Xavier Blanc, Combining independent model transformations, in: Proceedings of the 2010 ACM Symposium on Applied Computing, SAC '10, ACM, New York, NY, USA, 2010, pp. 2237–2243.

[80] EMF Views Project Homepage, http://atlanmod.github.io/emfviews/. (Accessed 2015).

[81] Dominique Blouin, Yvan Eustache, Jean-Philippe Diguet, Extensible global model management with meta-model subsets and model synchronization, in: Proceedings of the 2nd International Workshop on The Globalization of Modeling Languages co-located with ACM/IEEE 17th International Conference on Model Driven Engineering Languages and Systems, GEMOC@Models 2014, Valencia, 2014, pp. 43–52.

[82] Davide Di Ruscio, Ivano Malavolta, Henry Muccini, Patrizio Pelliccione, Alfonso Pierantonio, Model-driven techniques to enhance architectural languages interoperability, in: Juan de Lara, Andrea Zisman (Eds.), Fundamental Approaches to Software Engineering, Springer, Berlin, Heidelberg, 2012, pp. 26–42.

[83] Arnaud Blouin, Benoit Combemale, Benoît Baudry, Olivier Beaudoux, Kompren: modeling and generating model slicers, Software & Systems Modeling 14 (1) (2015) 321–337.

[84] Hugo Bruneliere, Jokin Garcia Perez, Manuel Wimmer, Jordi Cabot, EMF views: a view mechanism for integrating heterogeneous models, in: Paul Johannesson, Mong Li Lee, Stephen W. Liddle, Andreas L. Opdahl, Óscar Pastor López (Eds.), Conceptual Modeling, Springer International Publishing, Cham, 2015, pp. 317–325.

[85] Xavier Blanc, Alix Mougenot, Isabelle Mounier, Tom Mens, Incremental detection of model inconsistencies based on model operations, in: CAiSE '09: Proceedings of the 21st International Conference on Advanced Information Systems Engineering Amsterdam, The Netherlands, vol. 5565/2009, Berlin, Heidelberg, 8–12 June 2009, Springer Verlag, 2009, pp. 32–46.

[86] VIATRA Project Homepage, https://www.eclipse.org/viatra/. (Accessed 2020).

[87] Zoltan Ujhelyi, Gabor Bergmann, Abel Hegedus, Akos Horvath, Benedek Izso, Istvan Rath, Zoltan Szatmari, Daniel Varro, EMF-IncQuery: an integrated development environment for live model queries, in: Fifth issue of Experimental Software and Toolkits (EST): A Special Issue on Academics Modelling with Eclipse (ACME2012), Science of Computer Programming 98 (Part 1) (2015) 80–99.

[88] Alexander Egyed, Instant consistency checking for the UML, in: ICSE '06: Proceedings of the 28th International Conference on Software Engineering, Shanghai, China, 20–28 May 2006, pp. 381–390.

[89] Iris Groher, Alexander Reder, Alexander Egyed, Incremental consistency checking of dynamic constraints, in: David S. Rosenblum, Gabriele Taentzer (Eds.), Fundamental Approaches to Software Engineering, in: Lecture Notes in Computer Science, vol. 6013, Springer, Berlin, Heidelberg, 2010, pp. 203–217.

[90] Jordi Cabot, Ernest Teniente, Incremental evaluation of OCL constraints, in: CAiSE'06: 18th International Conference on Advanced Information Systems Engineering, Luxembourg, Luxembourg, 5–9 June 2006, in: Lecture Notes in Computer Science (LNCS), vol. 4001/2006, Springer Verlag, 2006, pp. 81–95.

[91] AM3 Project Homepage, https://wiki.eclipse.org/AM3, 2014.

[92] ATL Project Homepage, https://eclipse.org/atl/, 2015.

[93] Andrés Vignaga, Frédéric Jouault, María Cecilia Bastarrica, Hugo Brunelière, Typing artifacts in megamodeling, Software & Systems Modeling 12 (2013) 105–119.

[94] José E. Rivera, Daniel Ruiz-Gonzalez, Fernando Lopez-Romero, José Bautista, Antonio Vallecillo, Orchestrating ATL model transformations, in: Frédéric Jouault (Ed.), Proc. of MtATL 2009: 1st International Workshop on Model Transformation with ATL, Nantes, France, July 2009, pp. 34–46.

[95] ATLFlow Project Homepage, http://opensource.urszeidler.de/ATLflow/. (Accessed 2020).

[96] Moharram Challenger, Ken Vanherpen, Joachim Denil, Hans Vangheluwe, FTG+PM: Describing Engineering Processes in Multi-Paradigm Modelling, Springer International Publishing, Cham, 2020, pp. 259–271.

[97] AToMPM Project Homepage, https://atompm.github.io/. (Accessed 2020).

[98] AMW Project Homepage, https://wiki.eclipse.org/ATL/EMFTVM/. (Accessed 2020).

[99] Csaba Debreceni, Akos Horvath, Abel Hegedus, Zoltan Ujhelyi, Istvan Rath, Daniel Varro, Query-driven incremental synchronization of view models, in: Proceedings of the 2nd Workshop on View-Based, Aspect-Oriented and Orthographic Software Modelling, VAO '14, ACM, New York, NY, USA, 2014, pp. 31:31–31:38.

[100] Andreas Seibel, Regina Hebig, Stefan Neumann, Holger Giese, A dedicated language for context composition and execution of true black-box model transformations, in: 4th International Conference on Software Language Engineering (SLE 2011), Braga, Portugal, July 2011.

[101] Holger Giese, Stefan Neumann, Oliver Niggemann, Bernhard Schätz, Model-based integration, in: Holger Giese, Gabor Karsai, Edward Lee, Bernhard Rumpe, Bernhard Schätz (Eds.), Model-Based Engineering of Embedded Real-Time Systems - International. Revised Selected Papers, Dagstuhl Workshop, Dagstuhl Castle, Germany, November 4–9, 2007, in: Lecture Notes in Computer Science, vol. 6100, Springer, 2011, pp. 17–54.

[102] GME Project Homepage, http://www.isis.vanderbilt.edu/projects/gme/. (Accessed 2020).

[103] FUESD Project Homepage, http://www.adventiumlabs.com/our-work/products-services/fused-informational-video/, 2015.

[104] OSLC Project Homepage, http://open-services.net/. (Accessed 2020).

[105] Sebastian J.I. Herzig, Christiaan J.J. Paredis, Bayesian reasoning over models, in: 11th Workshop on Model Driven Engineering, Verification and Validation MoDeVVa 2014, 2014, p. 69.

[106] An Eclipse-based workbench for INTeractive Model Management, https://github.com/adisandro/MMINT, 2014.

[107] Alessio Di Sandro, Rick Salay, Michalis Famelis, Sahar Kokaly, Marsha Chechik, MMINT: a graphical tool for interactive model management, in: P&D@ MoDELS, 2015, pp. 16–19.

[108] MegaM@Rt2 framework, https://megamart2-ecsel.eu/, 2017.

[109] Wasif Afzal, Hugo Bruneliere, Davide Di Ruscio, Andrey Sadovykh, Silvia Mazzini, Eric Cariou, Dragos Truscan, Jordi Cabot, Abel Gómez, Jesús Gorroñogoitia, et al., The MegaM@Rt2 ECSEL project: megamodelling at runtime–scalable model-based framework for continuous development and runtime validation of complex systems, Microprocessors and Microsystems 61 (2018) 86–95.

[110] Regina Hebig, Andreas Seibel, Holger Giese, On the unification of megamodels, in: Vasco Amaral, Hans Vangheluwe, Cécile Hardebolle, Laszlo Lengyel, Tiziana Magaria, Julia Padberg, Gabriele Taentzer (Eds.), Proceedings of the 4th International Workshop on Multi-Paradigm Modeling (MPM 2010), in: Electronic Communications of the EASST, vol. 42, 2011.

[111] Dominique Blouin, Gilberto Ochoa Ruiz, Yvan Eustache, Jean-Philippe Diguet, Kaolin: a system-level AADL tool for FPGA design reuse, upgrade and migration, in: NASA/ESA International Conference on Adaptive Hardware and Systems (AHS), Montréal, Canada, June 2015.

[112] R. Hilliard, I. Malavolta, H. Muccini, P. Pelliccione, On the composition and reuse of viewpoints across architecture frameworks, in: 2012 Joint Working IEEE/IFIP Conference on Software Architecture and European Conference on Software Architecture, Aug 2012, pp. 131–140.

[113] Rich Hilliard, Ivano Malavolta, Henry Muccini, Patrizio Pelliccione, Realizing architecture frameworks through megamodelling techniques, in: Proceedings of the IEEE/ACM International Conference on Automated Software Engineering (ASE2010), ACM, New York, NY, USA, 2010, pp. 305–308.

[114] Multi-paradigm modeling for cyber-physical systems website, http://mpm4cps.eu/, 2020.

[115] Stefan Klikovits, Rima Al-Ali, Moussa Amrani, Ankica Barisic, Fernando Barros, Dominique Blouin, Etienne Borde, Didier Buchs, Holger Giese, Miguel Goulao, Mauro Iacono, Florin Leon, Eva Navarro, Patrizio Pelliccione, Ken Vanherpen, COST IC1404 WG1 Deliverable WG1.1: State-of-the-art on Current Formalisms used in Cyber-Physical Systems Development, Technical Report, 2020.

[116] Rima Al-Ali, Moussa Amrani, Soumyadip Bandyopadhyay, Ankica Barisic, Fernando Barros, Dominique Blouin, Ferhat Erata, Holger Giese, Mauro Iacono, Stefan Klikovits, Eva Navarro, Patrizio Pelliccione, Kuldar Taveter, Bedir Tekinerdogan, Ken Vanherpen, COST IC1404 WG1 Deliverable WG1.2: Framework to Relate / Combine Modeling Languages and Techniques, Technical Report, 2020.

[117] Jaroslav Keznikl, Tomas Bures, Frantisek Plasil, Ilias Gerostathopoulos, Petr Hnetynka, Nicklas Hoch, Design of ensemble-based component systems by invariant refinement, in: Proceedings of the 16th International ACM Sigsoft Symposium on Component-based Software Engineering, CBSE '13, ACM, New York, NY, USA, 2013, pp. 91–100.

[118] Michal Kit, Ilias Gerostathopoulos, Tomas Bures, Petr Hnetynka, Frantisek Plasil, An architecture framework for experimentations with self-adaptive cyber-physical systems, in: Proceedings of the 10th International Symposium on Software Engineering for Adaptive and Self-Managing Systems, SEAMS '15, IEEE Press, Piscataway, NJ, USA, 2015, pp. 93–96.

[119] Tomas Bures, Ilias Gerostathopoulos, Petr Hnetynka, Jaroslav Keznikl, Michal Kit, Frantisek Plasil, DEECo: an ensemble-based component system, in: Proceedings of the 16th International ACM Sigsoft Symposium on Component-based Software Engineering, CBSE '13, ACM, New York, NY, USA, 2013, pp. 81–90.

[120] Rima Al Ali, Tomas Bures, Ilias Gerostathopoulos, Petr Hnetynka, Jaroslav Keznikl, Michal Kit, Frantisek Plasil, DEECo: an ecosystem for cyber-physical systems, in: Companion Proceedings of the 36th International Conference on Software Engineering, ICSE Companion 2014, ACM, New York, NY, USA, 2014, pp. 610–611.

[121] Tomas Bures, Ilias Gerostathopoulos, Petr Hnetynka, Jaroslav Keznikl, Michal Kit, Frantisek Plasil, Gossiping Components for Cyber-Physical Systems, Springer International Publishing, Cham, 2014, pp. 250–266.

[122] F. Krijt, Z. Jiracek, T. Bures, P. Hnetynka, I. Gerostathopoulos, Intelligent ensembles – a declarative group description language and java framework, in: 2017 IEEE/ACM 12th International Symposium on Software Engineering for Adaptive and Self-Managing Systems (SEAMS), May 2017, pp. 116–122.

[123] Filip Krijt, Zbynek Jiracek, Tomas Bures, Petr Hnetynka, Frantisek Plasil, Automated dynamic formation of component ensembles - taking advantage of component cooperation locality, in: Proceedings of the 5th International Conference on Model-Driven Engineering and Software Development - vol. 1: MODELSWARD, INSTICC, SciTePress, 2017, pp. 561–568.

[124] Rima Al-Ali, Uncertainty-Aware Self-Adaptive Component Design in Cyber-Physical System, Technical Report D3S-TR-2019-02, Department of Distributed and Dependable Systems, Charles University, 2019.

[125] T. Bures, P. Hnetynka, J. Kofron, R.A. Ali, D. Skoda, Statistical approach to architecture modes in smart cyber physical systems, in: 2016 13th Working IEEE/IFIP Conference on Software Architecture (WICSA), April 2016, pp. 168–177.

[126] Bart Broekman, Edwin Notenboom, Testing Embedded Software, Addison Wesley, 2003.

[127] Sebastian Wätzoldt, Stefan Neumann, Falk Benke, Holger Giese, Integrated software development for embedded robotic systems, in: Itsuki Noda, Noriaki Ando, Davide Brugali, James Kuffner (Eds.), Proceedings of the 3rd International Conference on Simulation, Modeling, and Programming for Autonomous Robots (SIMPAR), in: Lecture Notes in Computer Science, vol. 7628, Springer, Berlin, Heidelberg, October 2012, pp. 335–348.

CHAPTER 5

An integrated ontology for multi-paradigm modelling for cyber-physical systems

Dominique Blouin[a]**, Rima Al-Ali**[b]**, Holger Giese**[c]**, Stefan Klikovits**[d]**,**
Soumyadip Bandyopadhyay[e]**, Ankica Barišić**[f] **and Ferhat Erata**[g]

[a]Telecom Paris, Institut Polytechnique de Paris, Paris, France
[b]Charles University, Prague, Czech Republic
[c]Hasso-Plattner-Institut, Potsdam, Germany
[d]University of Geneva, Carouge, Switzerland
[e]BITS Goa, Sancoale, India
[f]NOVA LINCS Research Center, Lisbon, Portugal
[g]Yale University, New Haven, CT, United States

5.1. Introduction

This chapter presents the MPM4CPS ontology that integrates the Shared, CPS and MPM ontologies by providing cross–cutting concepts between these domains. In particular, it formalises a new notion of viewpoint adapted from existing work. Such a viewpoint integrates the concepts of megamodel fragments of the MPM ontology with stakeholder concerns about the system of the Shared ontology and with the CPS constituent elements under development of the CPS ontology. The CPS ontology was specified in Chapter 3 in the form of a metamodel complemented with a feature model. After conversion of the provided CPS metamodel and feature model to an OWL ontology, we are able to refer to classes of the CPS ontology represented in OWL form from our MPM4CPS viewpoint notion.

We also extend the core notions of workflow processes of the Shared ontology to capture model-based development processes. In particular, we relate these processes to their employed viewpoints. Finally, we refine the notion of engineering paradigm introduced in the Shared ontology to define our notion of modelling paradigms, which adapts existing work to take into account our new viewpoint notion.

In the following, we first briefly introduce the state of the art that served as starting point to develop the ontology in Section 5.2. Then we detail the ontology that includes our extension of these state-of-the-art work in Section 5.3. Section 4.4 illustrates the usage of the ontology with the Ensemble-Based CPS and HPI CPSLab examples previously introduced in Chapter 2. We finally conclude this chapter and this first part of this book on foundations for MPM4CPS in Section 5.5.

Multi-Paradigm Modelling Approaches for Cyber-Physical Systems
https://doi.org/10.1016/B978-0-12-819105-7.00010-6

5.2. State of the art

Integration needs underlying the development of embedded real-time and cyber-physical systems have been outlined in [1]. These observations and state of the art work presented in this section will guide the development of the MPM4CPS ontology in order to cover the needs of the EBCPS and HPI CPSLab example development environments and their CPS case studies.

The MPM4CPS ontology is centred on a notion of viewpoint that relates parts of the CPS under study with the employed formalisms, languages and tools as informally defined in the framework of Broman et al. [2]. As already mentioned in Section 4.2 of Chapter 4, parts of the framework of Broman et al. also inspired the MPM ontology. Their framework is itself an adaptation of the ISO/IEEE standard 42010 standard on architecture descriptions [3], which we introduced in the Shared ontology presented in Chapter 2 except for its modelling-related notions that the MPM4CPS ontology redefines.

The MPM4CPS ontology enhances the framework of Broman et al. by enriching it with modelling notions of the MPM ontology such as megamodel and megamodel fragment. In addition, the ontology also considers model-based development processes to orchestrate the different modelling activities performed to develop specific parts of CPS using different megamodel fragments. Finally, the ontology also adapts an existing notion of modelling paradigms that characterises the formalism and development process of these development environments.

Therefore, we divide this state of the art along these three axes. The first part presents different existing works on viewpoint-related concepts including introducing the viewpoint concepts of the ISO/IEEE 42010 standard and that of the framework of Broman et al. [2]. The second part presents state of the art on the modelling of model-based development processes. Finally, the third part proposes a brief state of the art on modelling paradigms including the closely related topic of programming paradigms.

5.2.1 Viewpoints

We first complete the introduction of the ISO/IEEE 42010 standard by presenting its viewpoint-related notions. As depicted in Fig. 2.9, the ISO/IEEE 42010 standard defines its own notion of viewpoint, which associates a set of stakeholder concerns with a set of model kinds employed to address these concerns in the design of an architecture. These model kinds serve as typing elements for the models employed for specifying the architecture. Architecture viewpoints are then used to govern the view on the system according to the concerns and the model kinds.

The framework of Broman et al. [2] adapts the ISO/IEEE 42010 standard to CPS design by enriching the provided viewpoint notion. This is achieved by replacing the

Model Kind concept of the ISO/IEEE 42010 standard by the formalisms supporting the viewpoint to perform the different activities on the system models (Fig. 4.2). In addition, the Broman et al. viewpoint notion adds a link from the viewpoint to the parts of the system under design relevant to the concerns framed by the viewpoint. This is illustrated in Fig. 5.1 where the *Control Robustness Design*, *Control Performance Design* and *Software Design* example viewpoints are depicted for an advanced driver assistance system (ADAS) (e.g., adaptive cruise control) embedded control system. A matrix shows the parts of the system and their concerns along its two axes and cross–cutting points are defined and grouped to define viewpoints.

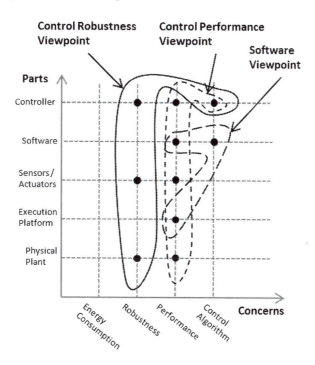

Figure 5.1 Example of Broman et al. viewpoints matrix (based on [2]).

For instance, the *Software Design* viewpoint frames the *Performance* and the *ADAS Algorithm* concerns, which are impacted by the *Software* and *Computing Platform* parts of the system. As can be seen in Fig. 5.2, such a viewpoint employs the state machines, dataflow and discrete event formalism to address these concerns in the CPS design.

These notions will be at the heart of our MPM4CPS ontology. Besides, we also want to relate these viewpoints to the development processes in which they are used, which is an important characteristic of CPS development approaches and environments. This is the topic of the next section.

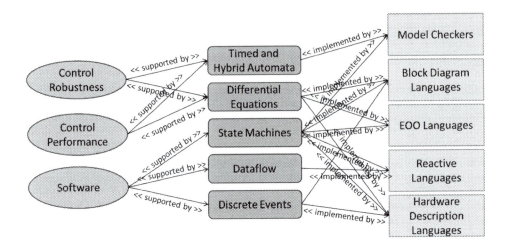

Figure 5.2 Example of viewpoints and their employed formalisms (based on [2]).

5.2.2 Model-based development process modelling

MPM advocates to model everything, at the most appropriate level(s) of abstraction, using the most appropriate formalisms, for the modelling activities to be performed. This includes not only the modelling of the system to be developed (at the micro modelling scale as defined by the MPM ontology) and the modelling of the models and modelling languages employed to develop the system (at the megamodelling scale), but also the modelling of model–based development processes that orchestrate the various activities performed on models of the system.

Unfortunately, we do not find many work on this topic in the literature. This is also the result of a systematic mapping study of MPM for CPS initiated during the MPM4CPS COST action [4] A main result of this study is that among the studies that reported a development process, 38% of them describe it informally often in a partial way; 30% describe it semi-formally, step by step, but still giving only natural language descriptions; and only 32% provide a formal model for specifying their process. Therefore, development process modelling remains an issue in MPM4CPS.

Nevertheless, we find the FTG+PM [5], which as already mentioned in the state of the art of Chapter 4, is the only model management language that formalises development processes. Therefore, it seems to be currently the most appropriate language to specify engineering processes in MPM. Fig. 5.3 shows the main concepts of the language. On the left hand side of the figure, we find the Formalism Transformation Graph (FTG) concepts, which describe a repository of transformation relations between formalisms. On the right hand side, we find the Process Model (PM) concepts. Those are based on the UML activity diagram, which has been adapted so that activities are typed by transformations defined in the FTG part. In addition, objects representing artifacts

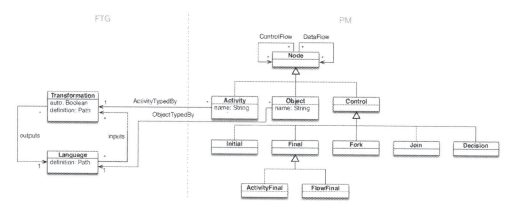

Figure 5.3 Overview of the FTG+PM concepts (from [5]).

processed by the activities are typed by languages of the FTG side therefore representing models taken as input by the activities when executing transformations.

The FTG part can be compared with the megamodel notion of the MPM ontology of Chapter 4. A difference, however, is that megamodels can be logically organised into fragments, which is not the case of the FTG. Besides, only one kind of relation is defined between formalisms, while the MPM ontology captures several kinds such as synchronisation, traceability and refinement relations. Finally, transformations relate formalisms and not modelling languages like for the megamodel relations of the MPM ontology. With our definitions of formalisms and modelling languages, we must relate modelling languages since they are concrete implementations of formalisms.

The FTG+PM languages, as well as the megamodelling notion of Chapter 4 and the workflow process concepts of the shared ontology of Chapter 2, will serve as basis for defining model-based process modelling in this MPM4CPS ontology.

5.2.3 Modelling paradigms

Another important MPM-related notion is that of a *modelling paradigm*. We initially found that there was no precise definition of this notion despite that it originated as early as 1996 [6]. However, the notion of modelling paradigm can be seen as a generalisation of the notion of *programming paradigm*, since programming languages can be seen as a subset of modelling languages. Programming paradigms, which originated as early as 35 years ago [7,8] have been defined to categorise the different approaches or styles used by the different programming languages. Due to the growing heterogeneity of software systems, which results in different kinds of problems to be solved, different approaches or paradigms to solve these problems had to be developed. Therefore, a plethora of programming languages has been created and categorised according to their underlying paradigms. However, the notion of programming paradigm, which varies from one author to another, was also never made very precise. The most precise definition that we find is that of [9], which informally defines a programming paradigm as "...*a set of programming concepts, organised into a simple core language called the paradigm's kernel language*". Yet, this definition remains vague.

Such lack of precise definition of programming and modelling paradigms was an important problem for our ontology of MPM, which triggered some work to precise the definition [10,11]. Part of this work has already been introduced in Chapter 2 that presented the Shared ontology and its paradigm subdomain. However, only the part that is independent of modelling was presented as its level of abstraction is adequate for modelling the more general notion of engineering paradigms.

We will build on the modelling-specific part of this work to precise the notion of modelling paradigm for this MPM4CPS ontology. As can be seed from Fig. 2.11, this work defines a modelling paradigm as a set of properties characterising the languages (including their semantics) and workflows (development processes) employed to develop systems. This can be seen as an enlargement of the notion of a programming paradigm, which typically only characterises programming languages and do not say anything about workflows. Examples of simple paradigms may be object orientation, which only pertains to formalisms, or agile development, which only characterises workflows. More complex paradigms examples discussed in the work of [11] are Synchronous Data Flow (SDF) and Discrete Event Dynamic Systems (DEv). These will be further discussed in Section 5.3 where we will introduce our MPM4CPS ontology in detail.

5.3. Ontology

In this section we provide an overview of the OWL MPM4CPS ontology that defines cross-cutting concepts between the Shared, MPM and CPS ontologies presented in the previous chapters. We first define the `MPM4CPSDC` domain concept class as a subclass of the `DomainConcept` class of the Shared ontology to organise the classes of the MPM4CPS integrating domain (Fig. 5.4). All classes defined in this section are said to be part of the MPM4CPS domain and will therefore be made subclasses of this MPM4CPS subdomain class.

Following the state of the art of the previous section, we first define our viewpoint notion inspired from the framework of Broman et al., but adapted for the megamodel and megamodel fragments of the MPM ontology. Then we refine the workflow process subdomain of the Shared ontology (see Section 2.5 of Chapter 2) to specialise it for model-based development. Finally, we refine the notion of engineering paradigms of the Shared ontology to define a more specific notion of modelling paradigm.

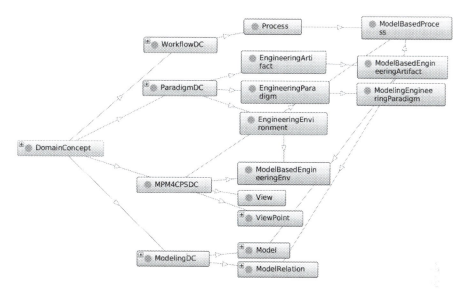

Figure 5.4 Overview of the OWL MPM4CPS ontology.

5.3.1 Viewpoint

We start by defining our notion of Viewpoint, which is inspired from the definition from the framework of Broman et al. (Fig. 4.2 and Fig. 5.1). Their definition is itself an adaptation of the notion of viewpoint from the IEEE 42010 standard [3] (Fig. 2.9). Compared to the notion of viewpoint from the IEEE 42010, the notion of the framework of Broman et al. replaces the ModelKind (Fig. 2.9) representing the modelling languages supporting a viewpoint by a set of Formalism.

However, we have seen in the MPM ontology how we defined the notion of MegamodelFragment that organises modelling languages conjointly used with their appropriate ModelRelations to support an activity of a development process (e.g. the Model Test activity of the simulation stage of the HPI CPSLab development process in Chapter 4). Therefore, we replace the set of formalism of the Broman et al. framework by a megamodel fragment, and we define the hasSupportingMegamodelFragments object property to relate a viewpoint to a set of supporting megamodel fragments.

According to the IEEE 42010 standard, a viewpoint *frames* the concerns of stakeholders (Fig. 2.9). Therefore we create the hasFramedConcerns object property to relate a viewpoint to a set of framed concerns as defined in the Shared ontology (Chapter 2).

In addition, a viewpoint *governs* an *architecture view*. We therefore define the View class and the hasGovernedView object property to relate a viewpoint to its view. Finally, we create another object property to relate a view to its employed models.

An addition of the framework of Broman et al. to the IEEE 42010 standard is to define a reference from a viewpoint to the system parts being developed using views of the viewpoint. We adopt this feature and create the `hasSystemConstituentElements` object property between a viewpoint and its system constituents or parts. The range of the property is the `ConstituentElement` OWL class, which was generated from feature of the feature model of the CPS ontology (Fig. 3.2).

We note that this makes our notion of viewpoint a cross–cutting concept between the MPM and CPS domains of our ontological framework. This explains why the viewpoint-related notions are part of the MPM4CPS integrating ontology, since they depend on all other ontologies. We also note that the link between a viewpoint and CPS constituent elements can be seen as being derived from the standard *representedBy* relation between a real system and its models [12].

This completes the definition of our viewpoint-related notions. We then consider how these viewpoints are used throughout development by introducing notions related to model-based workflows.

5.3.2 Model-based workflows

We have introduced core workflow process modelling notions inspired from the WFMC–TC-1025 standard in the Shared ontology of Chapter 2. On the other hand, the FTG+PM process modelling language introduced in Section 5.2.2 defines similar process modelling notions adapted from UML activity diagrams. An advantage of the WFMC–TC-1025 standard is its wider coverage of the workflow domain. For instance, embedded and external subprocesses can be modelled. Besides, the notion of activity performer is included to model tools or humans performing the activities. In FTG+PM the performer notion is implicitly embedded within the activity concept, the human / tool characteristic being declared in the transformation class (*auto* attribute of the *Transformation* class of Fig. 5.3).

Such richness of the workflow subdomain is required to model industrial-strength processes such as the one of the HPI CPSLab example, which reuses an existing methodology from the automotive domain (Fig. 2.19). However, as opposed to the FTG+PM language, the WFMC–TC-1025 standard does not say anything about modelling but only links activity performers to the processed data fields (Fig. 2.6), such data fields being typed by associated data types declared in processes. We already covered the FTG-related notions in our MPM ontology via megamodels capturing modelling languages and their relationships, but in a finer way by providing different kinds of transformation relations. Besides our megamodels are hierarchical[1] allowing to capture contextual model transformation relations.

Therefore, the process notions of our MPM4CPS ontology will reuse the core process notions of the Shared ontology, reuse the link between the process model and the FTG of the FTG+PM language, but replace the FTG by the richer megamodel

---

[1] This is not yet covered in this version of the MPM ontology.

fragment notion. In the following, we provide an overview of model-based workflow process notions as part of this MPM4CPS ontology.

5.3.2.1 Model-based process

We first subclass the `Process` class of the Shared ontology into the `ModelBasedProcess` class to represent all processes that manipulate models instead of data fields. The choice of the model-based name follows the naming of the model-based engineering paradigm.

As opposed to the WFMC-TC-1025 standard, a `ModelBasedProcess` does not declare the types of the data fields processed by activity performers. Instead, a set of viewpoints can be linked to a process through an object property `hasViewPoint`, so that data types are declared as the elements grouped under the associated megamodel fragment. Similar to the FTG+PM language, such types are modelling languages and their relations declared in the megamodel fragments.

5.3.2.2 Activity performers

The `ModelingTool` class of the MPM ontology was declared as a subclass of the `Tool` class of the Shared ontology. This class was also made a subclass of `Application` activity performer class of the workflow subdomain of the Shared ontology. Therefore, a modelling tool can be directly used as activity performer by any activity. Similarly, the `ModelingHuman` resource of the MPM ontology was also made a subclass of the `Application` activity performer class of the workflow subdomain. Therefore, modelling human can be directly used as activity performer for representing modelling activities that are performed manually.

Finally, the `hasTranformationSpecifications` object property of the MPM ontology can be used to relate an activity performer resource to the transformation relations it executes as declared in the activity's associated viewpoint.

5.3.3 Modelling paradigms

In the Shared ontology, we have provided the notion of `EngineeringParadigm` defined as a characterisation of environments in which engineering takes place. We refine this definition so that it characterises model-based engineering environments and their employed model-based artifacts.

In order to define our modelling paradigm notion, we first subclass the `EngineeringEnvironment` class of the Shared ontology by the `ModelBasedEngineeringEnv` class. Similarly, we define the `ModelBasedEngineeringArtifact` as a subclass of the `EngineeringArtifact`. We also make the `Model` and `ModelRelation` classes of the kernel model of the MPM ontology subclasses of `ModelBasedEngineeringArtifact`. Hence, all elements of the MPM ontology that are a subclass of `Model` are also `ModelBasedEngineeringArtifacts`. We also make the `ModelBasedProcess` class `ModelBasedEngineeringArtifacts` as well. Finally, we create the `hasModelingArtifacts` subproperty of the `hasArtifacts` object property having, respectively, the `ModelBasedEngineeringEnv` and `ModelBasedEngineeringArtifact` classes as domain and range.

We then define the `ModelingEngineeringParadigm` class as a subclass of the `EngineeringParadigm` class of the Shared ontology. As defined in the Shared ontology, an engineering paradigm declares characteristics of engineering artifacts that make the paradigm. Therefore in our refined definition of modelling paradigm, we refine the scope of the paradigm characteristics to modelling artifacts. Note that compared to the definition of [11], which was introduced in the state-of-the-art section, our definition is wider as its scope consisting of any modelling artifact is not restricted to modelling languages and workflows.

Several ways can be thought of to express the characteristics of a paradigm. In the Shared ontology, those characteristics are represented as properties including their expressions and decision procedures. We have reused these notions from the work of [11]. We therefore present in greater detail their way to express paradigm characteristics.

As can be seen from Fig. 2.11, the properties expressing characteristics are actually meant to be evaluated over a *paradigmatic structure*, which expresses some of the paradigm characteristics as patterns to be checked over formalisms and workflows. Therefore, the first step in checking that a set of modelling artifacts implements a paradigm is to match a paradigmatic structure over the considered modelling artifacts. Once a match is found, paradigmatic properties can be evaluated against the structure to completely determine if the artifacts satisfy the paradigm.

In order to express paradigmatic structures, the authors define an adaptation of the metamodel language where classes are placeholders to be matched by other languages. Therefore, modelling artifacts such as languages and processes must also be expressed as metamodels for paradigm satisfaction checking. The approach also relies on the fact that language semantics are expressed as semantic domains whose languages are also expressed as metamodels so that they can also be characterised by paradigmatic structures. In the approach, it is expected that providers of modelling artifacts are responsible to *map* their artifacts to the paradigmatic structure. This is one drawback of the approach, which may make the evaluation of paradigms satisfaction over industrial heterogeneous modelling environments difficult. For languages, workflows and their semantic domains already expressed as metamodels, this is relatively easier. However, for other metamodelling technical spaces such as grammars, a conversion is required. Besides, a language's semantics is often not formalised but embedded into programming code of the tool that executes the language, which makes the task of identifying paradigms even harder.

Nevertheless, the work of [11] is a first attempt and an ongoing work on defining modelling paradigms and is therefore subjected to be improved in the next future. Hence, at this stage, we will not yet precise in this ontology any approach for expressing paradigm characteristics. Besides, since the scope of our modelling paradigm definition is wider, the paradigmatic structure proposed in [11] would be too restrictive. Nevertheless, we will present an example of paradigm characteristics expression for the HPI CPSLab in Section 5.4.3.

Another important remark is the different definitions of the formalism notion between the authors of [11] and that of our MPM ontology. The authors use the definition of [13] (Fig. 4.1) where a formalism is defined as "*a language, a semantic domain and a semantic mapping function giving meaning to model in the language*". In the MPM ontology, we followed the framework of Broman et al. [2] where formalisms are "*mathematical objects consisting of an abstract syntax and a formal semantics*", and where modelling languages are "*concrete implementations of formalisms*". Broman et al. further note that a language's semantics may slightly deviate from the semantics of formalisms they realise.

The question is then to understand the difference between our definition of formalisms and the paradigms of [11], for those paradigms that only characterise formalisms (as per the definition of [13]). This is even more relevant as [11] use Synchronous Data Flow (SDF) and Discrete Event Dynamic Systems (DEv) as paradigms examples in their work. At some point, the authors even consider the SDF paradigm as a "*conceptual formalism*". In addition, these two paradigms are classified as formalisms in the Catalog of Formalisms, Modelling Languages and Tools [14], which was created during the MPM4CPS COST action project. An interesting future work would therefore consists of studying all other formalisms of the catalog such as Petri Nets, Abstract State Machines, Hybrid Automata using exploratory modelling to better understand the formalism notion.

5.4. Examples

In this section, we illustrate how the MPM4CPS integrated ontology presented in Section 5.3 covers the needs for a comprehensive modelling of development environments and their CPS case studies such as the EBCPS and HPI CPSLab examples. For each of these examples, we first start by modelling some of the employed viewpoints that we relate to their stakeholder concerns, the employed megamodel fragments of Chapter 4 and the parts of the CPS of Chapter 3.

Next, we show how the engineering methodologies of each example can be captured including their different stages and the activities employed at these stages. Each of these activities makes use of a workflow process that uses one of the previously defined viewpoints. The process decomposes the root activity into a sequence of subactivities whose activity performers are set as the proper modelling tools and humans described in Chapter 4. Note that we do not present a complete coverage of the engineering methodologies and processes here, but only the parts needed to illustrate the use of the MPM4CPS integrated ontology.

Finally, we illustrate the modelling of the Synchronous Data Flow (SDF) paradigm with the integrated ontology characterising one of the example development environment through some of its activities making use of modelling languages based on SDF.

5.4.1 EBCPS

We present the modelling details of the EBCPS example with regard to the MPM4CPS ontology. The EBCPS methodology is a simple methodology that contains only one stage. A more comprehensive example is provided for the HPI CPSLab methodology that contains three stages. In this example, we use the division of the megamodel fragments as defined in Chapter 4, which is per modelling language rather than per activity as for the case of the HPI CPSLab. Thus, our viewpoints can employ more than one megamodel fragment, and activity processes can refer to more than one viewpoint.

5.4.1.1 Methodology

We define the EBCPSMethodology as an instance of the `EngineeringMethodology` class, which contains a set of stages. In our case, the scenario of autonomous vehicle presented in Chapter 2 employs only one stage that consists of a simulation stage.

Methodology
- EngineeringMethodology: EBCPSMethodology
  - hasStages EngineeringStage: EBCPSSimulationStage

5.4.1.2 Methodology implementing process

Per our definition from the workflow subdomain of the shared ontology, a methodology does not specify how it is implemented since several implementations could exist for a given methodology. Therefore we define the EBCPSProcess as one implementation of the EBCPSMethodology.

This process defines an activity set that specifies activities and transitions to orchestrate the different root modelling activities of the EBCPSProcess. Each of these activities also implements the single simulation stage of the EBCPS methodology. The root activities for the simulation stage are:

ActivitySet
- hasSetActivities SubFlow: RequirementsRootActivity
  - isImplementingStage: EBCPSSimulationStage
  - hasSubProcess: RequirementsProcess
- hasSetActivities SubFlow: DesignRootActivity
  - isImplementingStage: EBCPSSimulationStage
  - hasSubProcess: DesignProcess
- hasSetActivities SubFlow: RuntimeRootActivity
  - isImplementingStage: EBCPSSimulationStage

    ⊛ hasSubProcess: RuntimeProcess
- hasTransitions Transition: Requirements2DesignTransition, Design2RuntimeTransition

The processes presented for each root activity declare finer grained activities such as editing a model, simulating a model, and checking simulation results. Some activities are performed by humans (i.e. a designer) such as in case of capturing the requirements and implementing the components with self-adaptation support. Other activities may be performed by tools such as simulation tools.

Each process uses a viewpoint that specifies the concerns addressed by the process, the part(s) of the CPS that are being developed and their employed megamodel fragment(s). These megamodel fragments specify the modelling languages and their relationships that support the process activities. We first present the viewpoints in the following.

Viewpoints
- Viewpoint: ComponentAutonomyVP
  - ⊛ hasFramedConcerns: SafetyConcern, AdaptabilityConcern, EfficiencyConcern
  - ⊛ hasSystemConstituentElements: Controller, Plant, Sensors, Actuators.
  - ⊛ hasSupportingMegaModelFragments: RequirementsMegaModelFragment, DesignMegaModelFragment, RuntimeMegaModelFragment, Self-Adaptation-MegaModelFragment
- Viewpoint: ComponentsCooperationVP
  - ⊛ hasFramedConcerns: SafetyConcern, EfficiencyConcern
  - ⊛ hasSystemConstituentElements: Communication, Controller, Plant, Sensors, Actuators
  - ⊛ hasSupportingMegaModelFragments: RequirementsMegaModelFragment, DesignMegaModelFragment, RuntimeMegaModelFragment, SimulationMegaModelFragment

Activity Subprocesses
- ModelBasedProcess: RequirementsProcess
  - ⊛ ActivitySet: editing model – To create an IRM model, the developer needs to define the invariants in the system and their relations to the processes and knowledge exchange, which can have assumptions. The processes are associated to a component role that is also part of the IRM model. After finishing the modelling part, the next activity is to generate the skeleton of the components and ensembles in Java code from the IRM model.
  - ⊛ hasViewPoint: ComponentAutonomyVP, ComponentsCooperationVP
- ModelBasedProcess: DesignProcess
  - ⊛ ActivitySet: editing model – Using the skeleton, the developer implements the processes in the component and the ensembles. In this part, the developer can

support mode-switching in the component. In the example, the vehicle component has a controller and a plant, which should be also implemented.
  ⁕ hasViewPoint: ComponentAutonomyVP, ComponentsCooperationVP
- ModelBasedProcess: RuntimeProcess
  ⁕ ActivitySet: running model, result model – The developer can execute the design model and have the components and ensembles running (i.e. DEECoRuntime-Model), which produces logs as outputs. The runtime environment monitors the component and performs the mode-switching when needed. At the same time, the simulation models run and synchronous with DEECoRuntimeModel.
  ⁕ hasViewPoint: ComponentControlVP, ComponentsCooperationVP

5.4.2 HPI CPSLab

The HPI CPSLab and its methodology including several stages supported by multi-formalisms settings provides a comprehensive example of a development process where the developed system is gradually built starting from pure models at the simulation stage and gradually integrating more and more of the real artifacts in the following prototyping and pre-production stages (Fig. 2.19).

We present the modelling of this process and of one of its employed modelling paradigms with the MPM4CPS ontology. We first present the modelling of the methodology and its stages. Then for each stage we present the detailed modelling of the different activities implementing the stage, including the employed viewpoints. Finally, we present the modelling of a simple modelling paradigm employed by the HPI CPSLab process and its viewpoints.

5.4.2.1 *Methodology*

We define the CPSLabMethodology instance of the `EngineeringMethodology` class of the workflow subdomain of shared ontology (Chapter 2) to capture the HPI CPSLab methodology and its set of stages as depicted in Fig. 2.19. This is achieved by representing each stage as an instance of the `EngineeringStage` class and creating instances of the `hasNextStage` object property defining the order between the stages.

Methodology
- EngineeringMethodology: CPSLabMethodology
  ⁕ hasStages EngineeringStage: `CPSLabSimulationStage`
    – hasNextStage: `CPSLabPrototypingStage`
  ⁕ hasStages EngineeringStage: `CPSLabPrototypingStage`
    – hasNextStage: `CPSLabPreproductionStage`
  ⁕ hasStages EngineeringStage: `CPSLabPreproductionStage`

5.4.2.2 Methodology implementing process

We define the root CPSLabProcess instance of the `ModelBasedProcess` class to implement the CPSLabMethodology. This process defines an activity set defining root activities and transitions to orchestrate them. Each of these activities contributes to implementing a stage of the EBCPS methodology. In addition, each root activity is decomposed into finer grained activities as declared with an associated subprocess.

ActivitySet
- hasSetActivities SubFlow: MTActivity
  * isImplementingStage: CPSLabSimulationStage
  * hasSubProcess: MTProcess
- hasSetActivities SubFlow: MiLActivity
  * isImplementingStage: CPSLabSimulationStage
  * hasSubProcess: MiLProcess
- hasSetActivities SubFlow: RPActivity
  * isImplementingStage: CPSLabSimulationStage
  * hasSubProcess: RPProcess
- hasSetActivities SubFlow: SiLActivity
  * isImplementingStage: CPSLabPrototypingStage
  * hasSubProcess: SiLProcess
- hasSetActivities SubFlow: HiLActivity
  * isImplementingStage: CPSLabPrototypingStage
  * hasSubProcess: HiLProcess
- hasSetActivities SubFlow: STActivity
  * isImplementingStage: CPSLabPreproductionStage
  * hasSubProcess: STProcess
- hasTransitions Transition: MT2MiLTransition
- hasTransitions Transition: MiL2RPTransition
- hasTransitions Transition: RP2SiLTransition
- hasTransitions Transition: MiL2SiLTransition
- hasTransitions Transition: SiL2HiLTransition
- hasTransitions Transition: HiL2STTransition

Transitions are instantiated with appropriate conditions (not presented here) to define the order of execution of activities. Note that the declared order must be overall consistent with the ordering of the implemented stages as defined by the methodology. Overall consistency means that activities of a stage that follows another stage should never be executed before the activities of the preceding stage have already been executed at least once. Indeed, although not shown in this case, root activity transitions may return back to an activity of a preceding stage in case the errors discovered during the current stage were introduced at an earlier stage.

In the next sections, for each stage we present the viewpoints and the root activity processes that employ them.

5.4.2.3 Simulation stage

The purpose of the simulation stage is to define the control laws of the system. As opposed to the two next stages, its activities only employ models as captured by the megamodel fragments of Chapter 4 to represent the system and its environment.

For the HPI CPSLab example, we define a set of viewpoints specific to each of the root activities. This differs from the EBCPS example where existing viewpoints (e.g. from a library) are reused to support the activities. In this case, we define three viewpoints to support each root modelling activity of the simulation stage as follows:

Viewpoints
- Viewpoint: CPSLabMTControlAlgorithmVP
  - hasFramedConcerns: ControlAlgorithm
  - hasSystemConstituentElements: Controller, Plant, Sensor, Actuator
  - hasSupportingMegaModelFragments: CPSLabMTMMF
- Viewpoint: CPSLabMiLControlAlgorithmVP
  - hasFramedConcerns: ControlAlgorithm, Stability, Safety, Reliability
  - hasSystemConstituentElements: Controller, Plant, Sensor, Actuator
  - hasSupportingMegaModelFragments: CPSLabMiLMMF
- Viewpoint: CPSLabRPControlAlgorithmVP
  - hasFramedConcerns: ControlAlgorithm, Stability, Safety, Reliability
  - hasSystemConstituentElements: Controller, Plant, Sensor, Actuator
  - hasSupportingMegaModelFragments: CPSLabRPMMF

Each of these viewpoints addresses the concern of the control algorithm of the system under design and is using a megamodel fragment defined in the example section of Chapter 4 of the MPM ontology to capture the employed modelling languages and their relations. In addition, the CPSLabMiLControlAlgorithmVP and CPSLabRPControlAlgorithmVP also address other concerns such as stability, safety, reliability, thanks to the plant model providing feedback to the controller as opposed to the static input data of the model test activity.

Each of these viewpoints describes a cyber-physical setting, at different levels of abstraction. The abstract control algorithm from the cyber domain captured by the MATLAB®/Simulink control model (ControlModel) is confronted with the physics as represented in the input data plus expected outcomes. This is model differently for each viewpoint as static data model (MT), plant model (MiL) and detailed robot model (RP). Therefore, all viewpoints cover the system constituents of interest represented by these models, which are the controller, plant, sensor and actuator elements. We further have a multi-formalism setting where the control is discrete while the input data is at least conceptually continuous.

Root Activity Subprocesses

Each root subflow activity is further described by a subprocess specifying its decomposition in terms of finer grained activities such as editing a model and executing a model transformation. Besides, the process, which is responsible for defining the context for executing its activities is associated with a viewpoint providing such context. We list below the root activity sub processes and their associated viewpoints. As an example, we present the fine grained activities of the model test subprocess in the next section.

- ModelBasedProcess: CPSLabMTProcess
  - ActivitySet: ...
  - hasViewPoint: CPSLabMTControlAlgorithmVP
- ModelBasedProcess: CPSLabMiLProcess
  - ActivitySet: ...
  - hasViewPoint: CPSLabMiLControlAlgorithmVP
- ModelBasedProcess: CPSLabRPProcess
  - ActivitySet: ...
  - hasViewPoint: CPSLabRPControlAlgorithmVP

Model Test Subprocess (CPSLabMTProcess)

We describe here as an example the set of subactivities that constitute the CPSLabMT-Process. Like for the process orchestrating root activities, this is achieved by creating a block activity and its activity set. But we first define activity performers to perform the activities.

Activity Performers

- ModelingHuman: ControlEngineerPerf
  - hasTranformationSpecifications: EditInputModelOperation, EditControlModel-Operation, EditPlantModelOperation, EditValidityResultsModel...
- ModelingTool: SimulinkTool
  - hasTranformationSpecifications: SimulateModelOperation, ...

Then we define the fine grained activities as per the list below.

ActivitySet

- hasSetActivities Activity: MTEditInputModel
  - hasActivityPerformer: ControlEngineerPerf
- hasSetActivities Activity: MTEditControlModel
  - hasActivityPerformer: ControlEngineerPerf
- hasSetActivities Activity: MTSimulateControlModel
  - hasActivityPerformer: SimulinkTool
- hasSetActivities Activity: MTCheckSimulationResults
  - hasActivityPerformer: ControlEngineerPerf
- hasTransitions Transition: EditInput2EditControlTransition

- hasTransitions Transition: EditControl2SimulateControlTransition
- hasTransitions Transition: CheckResults2EditControlTransition
  - hasCondition: ValidResults
- hasTransitions Transition: ...

Transitions are defined between the different activities. It should be noted that this is a simplified version of the real workflow as the order of the activities may depend on several conditions. For example, the transition CheckResults2EditControlTransition between the MTCheckSimulationResults and MTEditControlModel activities has a condition. Such condition evaluates some property of the ValidityResultsModel as was set by the designer during the MTCheckSimulationResults activity and indicating if the results are correct or not. If not correct, the control may be edited again. If correct, the process ends and by default returns to the calling subflow root activity.

At deployment, an application megamodel containing the models to be processed by activity performers can be bound to this structure for executing the process on real models.

The definition of the two other CPSLabMiLProcess and CPSLabRPProcess processes follows the same principles as that of the CPSLabMTProcess detailed above and is not presented. All details can be found in the ontology files accessible from [15].

5.4.2.4 *Prototyping stage*

Compared to the simulation stage, which only uses models, the focus of the prototyping stage changes from design to implementation. In this stage, the source code plays a major role and is gradually incorporated into the system under development. The purpose is to ensure that implementation constraints such as discretisation of variables and time due to the limited resources of the execution platform are handled appropriately to meet the system requirements. The concerns of this stage are therefore related to performance and accuracy and the activities of this stage are used to optimise related parameters such as data representation format, scheduling periods, sensor sampling rates, etc.

For this prototyping stage and the next pre-production stage of the HPI CPSLab methodology, we will only present the viewpoints specification. The modelling of root activity subprocesses is straightforward and follows the same principles as that of the simulation stage.

Like for the simulation stage, we define a viewpoint for each of the two root activities implementing the prototyping stage (Fig. 2.19). Therefore for each stage activity, we first present the activity and then define its supporting viewpoint.

Software in the Loop (SiL)

For the Software in the Loop (SiL) activity, the tool TargetLink, which is fully integrated into MATLAB Simulink, is used to generate C code from the Simulink behaviour model. This allows one to seamlessly migrate the functions and control algorithm from continuous behaviour of the model level to a discrete approximation implementation in software. Several parameters for code generation can be configured for the characteristics of the desired target platform. Different effects can be analysed and results can be compared to results obtained during the simulation stage.

The prototyping stage includes two forms of SiL activities. The first form consists of executing the developed software on a desktop computer against a simulator as depicted in Fig. 4.29. The detailed control algorithm from the cyber domain captured by the MATLAB/Simulink and AUTOSAR SystemDesk models (SystemModels) is joined with the physics as present in the sophisticated robot model of the simulator (RobotModel). Therefore, we clearly have a cyber-physical setting. We again have a multi-formalism setting as the control is discrete while the sophisticated robot mode is at least conceptually continuous. Consistency is checked via co-simulation as the software for the robot control runs in parallel with the sophisticated robot simulator.

The second form of SiL also consists of executing the software on a desktop computer but in this case against a real robot remotely controlled as depicted in Fig. 4.30. In this case, real sensor values are read from the robot and real actuators are controlled therefore including other effects such as sensor noise. The same detailed control algorithm from the cyber domain is this time joined with the physics of the real remotely controlled robot. Consistency is checked via co-execution as the software for the robot control runs in parallel with the remotely controlled robot.

Hardware in the Loop (HiL)

The hardware in the loop (HiL) activity consists of executing the software either on the robot as depicted in Fig. 4.30 or on special evaluation boards with debugging and calibration interfaces, which are similar to the final hardware execution platform. The almost unlimited execution resource of the desktop computer is replaced by the constrained resources of the final platform. Therefore concerns such as resources consumption could be added to this activity.

With its megamodel fragment, this activity ensures that the detailed control algorithm from the cyber domain captured by the MATLAB/Simulink model (ControlModel) is brought together with the physics as present in the robot and thus we have clearly a cyber-physical setting. Consistency is checked via executing the software on the robot.

All these SiL and HiL activities actually address similar concerns about the system. The difference resides in the fact that different models at different levels of abstraction (including real hardware) are used. Therefore we define the viewpoints of the following list:

Viewpoints
- Viewpoint: CPSLabSiLSoftwareDesignVP
  - hasFramedConcerns: ControlAlgorithm, Stability, Safety, Reliability

- ◦ hasSystemConstituentElements: Software (Cyber in feature model), Execution-Platform (Control in feature model)
  - ◦ hasSupportingMegaModelFragments: CPSLabSiLMMF
- • Viewpoint: CPSLabHiLSoftwareDesignVP
  - ◦ hasFramedConcerns: ControlAlgorithm, Stability, Safety, Reliability, Resources Consumption
  - ◦ hasSystemConstituentElements: Software (Cyber in feature model), Execution-Platform (Control in feature model)
  - ◦ hasSupportingMegaModelFragments: CPSLabHiLMMF

5.4.3 Modelling paradigms

We present here an example the modelling of one of the paradigms employed by the HPI CPSLab, which is Synchronous Data Flow (SDF), following its description in [11]. Then we present the modelling of the overall HPI CPSLab model-based development environment employing this paradigm within the MATLAB/Simulink tool captured in its megamodel.

SDF is a special case of the Data Flow paradigm [16]. It specifies a directed graph of computations nodes (also called *blocks*) exchanging signals representing infinite streams of data. Computation units execute whenever input data becomes available. Such units without input can fire at any time. They can be atomic or composite by encapsulating a subgraph. Arcs connect nodes together and describe how streams of data flow through the computation nodes. Execution consists of accumulating enough samples within the system as produced by blocks without inputs and performing the subsequent nodes computations thus consuming sample data on inputs and concurrently producing outputs.

The SDF Paradigm [17] is a specialisation of Data Flow where all computation nodes are *synchronous*, meaning that each block explicitly defines how many samples are consumed and produced. In their work, [11] describe the SDF paradigm as exhibiting the following characteristics:

- • SignalProperty: Signals composed of an infinite ordered stream of Samples are present.
- • DirectedGraphProperty: A *directed* graph with Blocks as nodes and Arcs is present.
- • BlocksPortsProperty: Blocks possess Ports that explicitly define how many Samples are used (consumed by Inputs, or produced by Outputs).
- • ArcsProperty: Arcs connect Ports and instantaneously transmit Signals. Note that a Port may be plugged to several Arcs but shortcuts are prohibited. Arcs are forbidden to connect as source and target Ports of the same Type.
- • MemoryFullProperty: A MemoryFull Block should always define an extra Port corresponding to initial conditions.

Following this, we define an engineering paradigm to represent SDF as follows:

Engineering Paradigms
- ModelingEngineeringParadigm: SDFParadigm
  - hasCharacteristics: SDFParadigmCharacteristics
    - hasProperties: SignalProperty, DirectedGraphProperty, BlocksPortsProperty, ArcsProperty, MemoryFullProperty

There are several ways such properties could be specified. If all languages and their semantics expressed as semantic domains are encoded using the same technical space such as Ecore metamodels, then these properties could be encoded as graph patterns using tools such as Henshin[2] or SDM.[3]

It should be noted that the aforementioned Catalog of Formalisms, Modelling Languages and Tools already declares SDF as being a formalism and not a paradigm. Besides, several modelling languages of the catalog such as Simulink are declared as being based on SDF. Therefore, the question of whether SDF should be classified as a formalism or a paradigm remains. A deeper study of the catalog's formalisms could help answer that question.

Next, we model the HPI CPSLab model-based engineering environment, which uses the CPSLabMM megamodel presented in the HPI CPSLab example of Chapter 4 and the CPSLabProcess development process defined in the previous section:

Engineering Environment
- ModelBasedEngineeringEnv: CPSLabEngineeringEnv
  - hasModelingArtifacts: CPSLabMM, CPSLabProcess
  - isBasedOnParadigms: SDFParadigm

It should be noted that the `isBasedOnParadigms` property at the engineering environment level is derived from the same property of Simulink language captured in the mega model. This can be determined from the SDFParadigm and its properties evaluated over the language and its semantics.

5.5. Conclusion

This chapter presented an integrated ontology for MPM4CPS that captures cross-cutting concepts between the Shared, CPS and MPM ontologies, respectively, presented in Chapters 2, 3 and 4. It defines notions such as model-based development processes, their employed viewpoints supported by megamodel fragments of the MPM ontology and the CPS parts under development covered by these viewpoints and defined in the CPS ontology. It finally introduced modelling paradigm notions at the heart of MPM4CPS as a refinement of the more general notions of engineering paradigms de-

[2] https://www.eclipse.org/henshin/.

[3] https://www.hpi.uni-potsdam.de/giese/public/mdelab/mdelab-projects/story-diagram-tools/.

fine in the Shared ontology. All these elements are captured under the model-based engineering environment concept defined in this integrating ontology.

As much as possible, these notions of the ontological framework have been grounded on existing work and standards such as the Workflow Management Coalition (WfMC) WFMC-TC-1025 [18] and the IEEE 42010 [3] standards, which have been extended / adapted for the MPM4CPS domains. Besides benefiting from the maturity of this work, this also allows stakeholders already familiar with these standards to understand and use the framework with less effort.

In addition, the adopted exploratory modelling approach based on the characterisation of existing development settings such as the EBCPS and HPI CPSLab with their CPSs case studies triggered several adjustments of the ontologies to account for existing setups. For instance, the comparison of the modelling of the two examples shows that the framework needs to be flexible enough to be able to capture the heterogeneous practices of industry. Despite that both examples come from the academic world, we have seen that yet they do not organise their megamodel fragments according to the same criterion; per language for the EBCPS and per root process activity for the HPI CPSLab. Besides, viewpoints can be constructed with different objectives such as those of the EBCPS that pre-exist the development process and that must be used as is, while for the HPI CPSLab, the viewpoints are specially constructed to support specific process activities. An even larger heterogeneity can be expected from legacy industrial settings and therefore, being able to cover a large spectrum of practices is essential for this framework to be useful for industry, otherwise disrupting existing industrial settings to adjust to the framework would limit its adoption.

Like for the case of biological science, the classification proposed in this work is not final and will evolve as new MPM4CPS environments are discovered. In particular, the notion of engineering paradigm, its modelling paradigm specialisation, the notion of formalism versus paradigm and in multi-paradigm modelling further needs to be investigated, first to be able to discover and understand paradigms from existing development settings and second to support a constructive way of building new MPM4CPS engineering settings based on a set of given paradigms. We hope that our framework can form a solid foundation for implementing a model management solution to relate and combine modelling languages and tools supporting MPM4CPS, as per the original goal of the MPM4CPS COST action project. This will be considered in future work where constructive modelling will be used to build the envisaged solution using the ontology as foundation.

References

[1] Holger Giese, Stefan Neumann, Oliver Niggemann, Bernhard Schätz, Model-based integration, in: Holger Giese, Gabor Karsai, Edward Lee, Bernhard Rumpe, Bernhard Schätz (Eds.), Model-Based Engineering of Embedded Real-Time Systems - International. Revised Selected Papers, Dagstuhl

Workshop, Dagstuhl Castle, Germany, November 4–9, 2007, in: Lecture Notes in Computer Science, vol. 6100, Springer, 2011, pp. 17–54.

[2] David Broman, Edward A. Lee, Stavros Tripakis, Martin Törngren, Viewpoints, formalisms, languages, and tools for cyber-physical systems, in: Proceedings of the 6th International Workshop on Multi-Paradigm Modeling, MPM '12, ACM, New York, NY, USA, 2012, pp. 49–54.

[3] ISO/IEC/IEEE 42010:2011. Systems and software engineering - architecture description, the latest edition of the original IEEE std 1471:2000, recommended practice for architectural description of software-intensive systems, 2011.

[4] Ankica Barišić, Dušan Savić, Rima Al-Ali, Ivan Ruchkin, Dominique Blouin, Antonio Cicchetti, Raheleh Eslampanah, Oksana Nikiforova, Mustafa Abshir, Moharram Challenger, Claudio Gomes, Ferhat Erata, Bedir Tekinerdogan, Vasco Amaral, Miguel Goulao, Systematic Literature Review on Multi-Paradigm Modeling for Cyber-Physical Systems, December 2018.

[5] Moharram Challenger, Ken Vanherpen, Joachim Denil, Hans Vangheluwe, FTG+PM: Describing Engineering Processes in Multi-Paradigm Modelling, Springer International Publishing, Cham, 2020, pp. 259–271.

[6] Hans Vangheluwe, Ghislain Vansteenkiste, Eugene Kerckhoffs, Simulation for the future: progress of the ESPRIT Basic Research working group 8467, in: European Simulation Symposium (ESS), SCS, 1996.

[7] Brent Hailpern, Guest editor's introduction multiparadigm languages and environments, IEEE Software 3 (01) (jan 1986) 6–9.

[8] Pamela Zave, A compositional approach to multiparadigm programming, IEEE Software 6 (05) (sep 1989) 15–18.

[9] Peter Van Roy, Concepts, Techniques, and Models of Computer Programming, MIT Press, 2012, pp. 9–47, chapter Programming Paradigms for Dummies: What Every Programmer Should Know.

[10] M. Amrani, D. Blouin, R. Heinrich, A. Rensink, H. Vangheluwe, A. Wortmann, Towards a formal specification of multi-paradigm modelling, in: 2019 ACM/IEEE 22nd International Conference on Model Driven Engineering Languages and Systems Companion (MODELS-C), 2019, pp. 419–424.

[11] M. Amrani, D. Blouin, R. Heinrich, A. Rensink, H. Vangheluwe, A. Wortmann, Multi-paradigm modeling for cyber-physical systems: a descriptive framework, International Journal on Software and Systems Modeling (SoSyM), in press.

[12] Jean Bézivin, On the unification power of models, Software and Systems Modeling 4 (2) (2005) 171–188.

[13] Holger Giese, Tihamer Levendovszky, Hans Vangheluwe (Eds.), Summary of the Workshop on Multi-Modelling Paradigms: Concepts and Tools, 2006.

[14] Stefan Klikovits, Rima Al-Ali, Moussa Amrani, Ankica Barisic, Fernando Barros, Dominique Blouin, Etienne Borde, Didier Buchs, Holger Giese, Miguel Goulao, Mauro Iacono, Florin Leon, Eva Navarro, Patrizio Pelliccione, Ken Vanherpen, COST IC1404 WG1 Deliverable WG1.1: State-of-the-art on Current Formalisms used in Cyber-Physical Systems Development, Technical Report, 2020.

[15] Multi-paradigm modeling for cyber-physical systems website, http://mpm4cps.eu/, 2020.

[16] Ian Watson, John G. Gürd, A practical data flow computer, IEEE Computer 15 (1982) 51–57.

[17] Edward A. Lee, David G. Messerschmitt, Static scheduling of synchronous data flow programs for digital signal processing, IEEE Transactions on Computers 36 (1) (1987) 24–35.

[18] WFMC-TC-1025 Workflow Management Coalition Workflow Standard, Process Definition Interface – XML Process Definition Language, 2005.

PART 2

Methods and tools

CHAPTER 6

Enabling composition of cyber-physical systems with the two-hemisphere model-driven approach

Oksana Nikiforova[a], **Mauro Iacono**[b], **Nisrine El Marzouki**[c], **Andrejs Romanovs**[a] **and Hans Vangheluwe**[d]

[a]Riga Technical University, Riga, Latvia
[b]University of Campania "Luigi Vanvitelli", Caserta, Italy
[c]LIMS Laboratory, USMBA, Fez, Morocco
[d]University of Antwerp and Flanders Make, Antwerp, Belgium

6.1. Introduction

Cyber-physical systems (CPS) have never been more central to the corporate strategy than today. The features they offer, reliability, performance and robustness, are some of the queen qualities that allow companies to be competitive. To cope with the complexity of the execution of such heterogeneous systems it is necessary to define an approach to tame its complexity [1]. A suitable approach should be flexible and generic in order to adapt to any type of component of such systems, thus it should offer the ability to manage system composition [2].

The two-hemisphere model-driven (2HMD) approach [3] has been successfully applied for domain modelling and software design [4]. One of the most distinguished features of this model is its applicability for both human understanding and automatic transformation. In this chapter we illustrate how the 2HMD approach may be applied to model and compose CPS.

The goal of this chapter is to show how the problem of complex system composition from smaller parts can be solved by using the 2HMD approach for modelling CPS components. From the point of view of the 2HMD approach, each component of a CPS may be considered as a conceptual class, which performs particular operations and meets the defined requirements. The requirements are derived from the model that consists of functional and conceptual "hemispheres". The 2HMD approach is thus applicable for both modelling components and the process to be supported by those components at the same level of abstraction. Moreover, the 2HMD approach can help to identify conflict situations, where the additional analysis is required for sharing responsibilities between system components.

Components of cyber-physical systems are presented in Section 6.2. Features of cyber-physical systems are discussed in the context of the necessity to model and compose them in Section 6.3. The essence of the two-hemisphere model-driven approach is clarified in Section 6.4. Application of 2HMD approach for composition and modelling of components of CPS are outlined in Section 6.5. Conclusions are given in Section 6.6.

6.2. Components of cyber-physical systems

The significant increase of use of the information and communication technologies in public and private sectors of economics in the last decade radically changed the way and shape in which humans observe and control the world. The rise of digital society opens new opportunities to interact with other persons, physical and virtual objects, uniting all of them into a new domain that has been named cyber-physical systems (CPS) – specialised computing systems that interact with control or management objects, integrating computing, communication, data processing and storage with real world objects and physical processes. All above-mentioned processes must occur in a real-time, safe, secure and efficient manner. Cyber-physical systems must be scalable, cost-effective and adaptive [1]. In CPS, physical and software components are deeply intertwined, each operating on different spatial and temporal scales, exhibiting multiple and distinct behavioural modalities, and interacting with each other in a myriad of ways that change with context [5]. In this chapter, we will present an overview of the CPS nature, main components and characteristics.

The growing need for information management systems to support different application fields is driven by an increase of both the information intensity and the information content of artifacts and related processes. An information management system is a computing system designed for management purposes, in most cases separated from the controlled object [6]. Most of the currently used information management systems are embedded systems and networked systems, closely related to the control or management objects. The availability of smaller, cheaper and richer microprocessor-based devices led to the possibility of an increase of information about controlled objects and to the integration of a significant part of the related portion of the information management system (its management object) into the controlled object. System elements are gradually becoming cheaper and their integration increases, as well as the security level and the opportunity to combine them in controlled networks. A downturn in embedded systems element prices and increasing connection with physical management objects led to the appearance of cyber-physical systems, in which rich information-related and control-related elements may be incorporated into controlled objects.

To frame CPS and their roots and evolution with respect to computing systems, the classification proposed by David Patterson and John Hennessy [7] provides a good guidance. They classified systems by their area of use, categorising computing systems into three main families, one of which of our interest, namely "embedded systems" (the others being "desktop computers" and "servers"), which have been further divided into:

- automatic control systems;
- measuring systems and systems that read information from sensors;
- real-time "question – answer" type information systems;
- digital data transmission systems;
- complex real-time systems;
  moving objects management systems;
- subsystems of a general purpose computer system;
- multimedia systems.

The concept of embedded systems appeared in the early 50s; and continues its development still today. The main key steps of the evolution of embedded systems have been: (a) information management systems, 1960s, (b) embedded computing systems, 1970s, (c) embedded distributed systems, 1990s, and finally (d) cyber-physical systems, since 2006 [8].

The first definition of the term "cyber-physical system" was authored by Dr. Helen Gill of the National Science Foundation, USA, in 2006. She defined CPS as "physical, biological, and engineered systems whose operations are integrated, monitored, and/or controlled by a computational core. Components are networked at every scale. Computing is deeply embedded into every physical component, possibly even into materials. The computational core is an embedded system, usually demands real-time response, and is most often distributed" [9].

Modern CPS can be viewed as an umbrella framework, focused on the fundamental intersection of computation, communications and physical processes without suggesting any particular implementation or application, and encompassing many other technologies, connecting the physical world with the cyber world [10]. First, CPS encompass industrial control systems, such as SCADA Systems, Distributed Control Systems and Programmable Logic Controllers. Other technologies belonging to the CPS domain are the ones related to Industry 4.0, Internet of Things and Internet of Everything, Machine-to-Machine (M2M), different "smart" applications, Fog Computing and TSensors/trillion sensors [11]. Wireless sensor networks play an important role in CPS as their ability for a capillary collection of physical data it is one of the main driving factors of distributed CPS applications. CPS typically consist of sensors and processors attached to key control points, monitoring their status, pressure, temperature, flow, etc., impacting the CPS operations. If necessary, CPS processors may automatically adjust the equipment to keep its operation conditions as intended, with monitoring, processing and adjustments usually occurring with no human intervention, while the opportunity for manual intervention still exists in CPS, e.g. in case of reconfigurations or regime/goal changes. Local processing units send processed data to the CPS information management systems that use them for efficiency analysis, trends detection or diagnostic needs. Cloud computing is considered as an integral part of these data processing operations [12].

The concept map of CPS, developed by Berkeley University [13] shows the complexity of real physical processes and virtual computational processes integration (see Fig. 6.1).

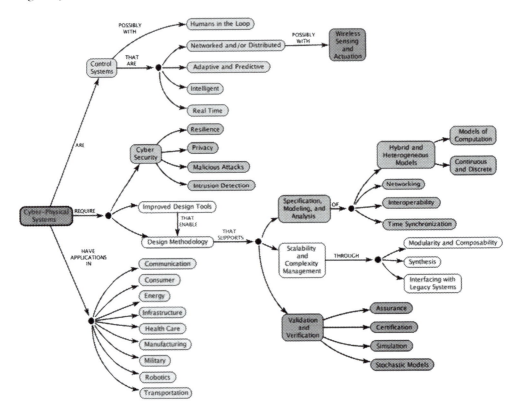

Figure 6.1 A concept map of cyber-physical systems [13].

Another CPS definition, from www.cpsweek.org, pointed out the complexity of CPS too: "CPS are complex engineering systems that rely on the integration of physical, computation, and communication processes to function". Compared with embedded systems, much more physical components are involved in CPS. In embedded systems, the key focus is on the computing elements, but in CPS it is on the link between computational and physical elements: in their essence, CPS parts exchange information with each other, which is why the communication component is so crucial: for this reason, CPS may be considered C3 (Computation, Communication and Control) systems. Links improvement between computational and physical elements may extend CPS usage possibilities [1].

CPS are used in multiple areas listed as "Have applications in" in Fig. 6.1. In all cases, increased use of CPS is closely related with cyber and IT risks that need to be managed appropriately [14], specially if considering the great influence of effective risk management on profit abilities in modern business systems, especially highly automated ones with complex use of Information Technology. IT risk consists not only of breakdowns in computer software or hardware, or lack of expertise of the IT staff. IT risk also may be the risk of loss that originates from computer software malfunction, such as a manufacturer software license expiration or glitches, and the ways it affects corporate activities. It may relate to risk of loss resulting from theft of company data or client information too [15].

6.3. Cyber-physical systems in the context of system composition

The nature of CPS involves different disciplines, such as cybernetics, mechatronics, design and process science [9,16,17]: as a consequence, a fundamental characteristic of CPS is their complexity, in terms of requirements, specification, verification, interactions with the physical part and constraint management: this makes analysis of CPS a difficult task that is still for many reasons an open issue in the research domain. For non-trivial CPS examples, one of the relevant complexity factors is that the coupling between the (continuous) dynamics of the main physical part of the system and the software related (discrete or event-based) evolution of the interacting computer part of the system may generate additional emergent dynamics locally and globally in the system, with consequent effects on different levels of abstractions or scale in the system. The system nature of the computer-based part adds further complexity elements in terms of timeliness or dependability that might affect the stability of the behaviours of the overall system.

With respect to other application fields that involve the external environment as either active or interactive context, e.g. the Internet of Things (IoT), in CPS physical and computational components have a higher level of interdependency and interaction, raising the need for a deeper understanding of the problems that arise due to the interaction between time continuous physical components and external phenomena and event-based computational components [18]. Moreover, the dynamic interaction with the environment poses an evolutionary aspect in CPS, which may be seen in terms of system intelligence [19,20] and of possibly limited knowledge about the external environment and its state, which should be managed to ensure that system requirements in terms of stability, safety, performance, and timeliness, are anyway ensured.

The design strategy of CPS is of course influenced by the coexistence of the different domains in a single system that fosters compositional methods. Design by composition allows, first, focusing on a single domain for components that are not at the boundary between domains, and exploiting well–established, conventional tools and practices, and, second, guides a proper identification and delimitation of responsibility of the components that have to constitute the border, and should be compliant to specific inter-domain constraints. Consequently, designers are suggested how to obtain a correct specification and implementation of the expected (emergent) system level behaviours by localising in the architecture the actual interactions and exchanges between domains, lowering the complexity of the design activities. Conversely, analysis of existing CPS or design of specifications benefits of decomposition approaches too, exploiting level-based processes. In any case, the heterogeneity of components and the different abstraction level and focus at each level requires different professional experiences and heterogeneous design and analysis tools and techniques, thus multiformalism or multiparadigm approaches may provide a profitable solution. The need for new concepts for design is discussed in [21,22], while [23] discusses the system development process.

Multiformalism [24] and multiparadigm [25] approaches are based on a process that defines a model by composing or deriving or generating submodels or auxiliary models that use different modelling languages to benefit of specific solutions. This may happen by means of closed or extensible modelling frameworks, which in general provide a predefined set of modelling formalisms or allow the definition of custom modelling formalisms, and coordinate the definition of complex models (e.g. by means of metamodelling, orchestration, transformation, oriented to functional or non-functional specification or to analysis). An example of an extensible multiformalism approach is provided by SIMTHESys [26], which offers the integration between domains of CPS in terms of hybrid systems [27] and decoupling of state spaces to lower the complexity of the analysis [28]; another example of more specific approach, based on decomposition or composition, is Ensemble Component-Based Systems [21,29]. The composition of CPS, considering CPS as systems of systems, is examined in [30] and [31], which stress the importance of architecture as a main support to vehicle the needs of the design process and to provide an infrastructure that allows compositional approaches by encapsulating cyber and physical aspects and integrating, at a higher level of abstraction, CPS concerns. Finally, the two-hemisphere model, that is, the composition approach used in this chapter, has been introduced in [3] and is here detailed and extended. It promotes working on high-level design abstractions to guide the definition of a conceptual prototype as a means to automatically compose models defined in terms of class diagrams.

CPS are dynamic systems in which components, according to the current context, are composed or decomposed. Modelling of these systems is based on submodels of the architectural and the behavioural parts that together provide a means for the analysis and verification of their overall requirements. The modelling approach supports decomposition of their structure and is founded on representing components as black boxes inside the system, which feature a finite set of roles, each of which is a communication interface towards other components. Roles are defined in turn as "knowledge", in terms of a set of attributes (including an enumeration of possible modes, and the references to its sub-components), and a mode-switch table (that can access knowledge, thus can define component behaviour in terms of context constraints with respect to modes) [32].

A mode corresponds to a set of processes that are executed when it is active on the component: consequently, each mode-switch table of the model represents the behaviour of the role of a component, and resumes the activity flow in a component. Besides this structural decomposition description, the decomposition process is described by "ensembles" that group three kinds of information: 1) the roles played by each component and its current mode; 2) the context, expressed in terms of membership conditions; 3) the exchange of knowledge between nodes that is the behaviour of an ensemble in terms of constraints [33].

In [34] and [32] the problem of managing system uncertainty is considered, in terms of an adaptation process that encompasses it when modelling both physical and computational elements: e.g., the case of an Ordinary Differential Equations (ODE)-based description is presented in [34], to account for delays consequences on physical components behaviour caused both by uncertain conditions in communications or in computing processes. A similar case is examined in [32] about precision of sensing and its management by means of a self-adaptation process based on an extension of mode-switch logic to include statistical testing performed on historical data: in this case, confidence levels in mode-switching conditions with short time prediction are targeted, and uncertainty is explicitly represented in the architectural view to leverage with a small additional effort existing analysis techniques and tools.

Finally, composition and decomposition in general purpose suitable abstractions, such as multiagent systems plays an important role in terms of global consequences of local behaviour in terms of emergent system behaviour: relevant, widely analysed application areas are social system models, traffic dynamics, complex business processes; an example of tool that can be adopted for such problems is Role-Activity Diagrams [35], in which, again, the notion of role with a local state-based evolution is used to describe the behaviour of a set of role instances, and the notion of actor represents entities in a business process that enact role instances, which in turn perform (role) activities, transferring control from a state to another in the role and describing interactions in terms of activities that happen in coordination between different roles. Ref. [36] shows how workflows that conform to these characteristics can be formally modelled as a knowledge-based business agent architecture, thus can be simulated and executed by means of specialised agent-oriented languages (e.g., Jason), or by common functional languages (e.g., F# [37]).

6.4. Two-hemisphere model-driven approach

The variety of modelling capabilities and the ability to express links traceability are decisive assets to manage system complexity. The transformation tool takes one model as input and produces a second model as its output. The two-hemisphere model-driven approach [3] proposes the use of business process and concept modelling to represent systems in a platform independent manner and describes how to transform those models into UML diagrams. This strategy was proposed for the first time in [38], where the general framework for object-oriented software development was presented and the idea of the usage of two interrelated models for software system development was stated and discussed.

The title of the proposed strategy is derived from cognitive psychology [39]. The human brain consists of two hemispheres: one is responsible for logic and the other for concepts. Harmonic interrelated functioning of both hemispheres is a precondition of an adequate human behaviour. A metaphor of two hemispheres may be applied to software development process because this process is based on investigation of two fundamental things: business and application domain logic (processes) and business and application domain concepts and relations between them. The two hemisphere approach proposes to start the process of software development based on a two hemisphere problem domain model, where one model reflects functional (procedural) aspects of the business and software system, and the other model reflects the corresponding concept structures. The co-existence and inter-relatedness of these models enables the use of knowledge transfer from one model to another, as well as the utilisation of particular knowledge completeness and consistency checks [3]. Fig. 6.2 shows the essence of a two-hemisphere model-driven approach.

The schema presents how elements of business process model (graph G1 in Fig. 6.2) and concept model (graph G2 in Fig. 6.2) are transformed into elements of a UML communication diagram (graph G4 in Fig. 6.2) and then into a class diagram (graph G5 in Fig. 6.2). A notation of the business process model (graph G1 in Fig. 6.1), which reflects functional perspectives of the problem and application domains, is optional. However, it must reflect business processes and information flows, and information flows have to be typed by corresponding concepts in the concept model (graph G2 in Fig. 6.2), which is used in parallel with business process model. A concept model is a variation of well-known entity relationship (ER) diagram notation [40] and consists of concepts (i.e. entities or objects) and their attributes. The original notational conventions proposed by the two-hemisphere model give a possibility to address concepts in concept model to information flows in a process model, thus making two models co-related [4].

The transformations of graph G1 to G4 using an intermediate model (G3) are shown in Fig. 6.3. The process model of two-hemisphere model is represented as a graph G1 (P, U), where P = {P1, P2, P3, P4, P5} is the set of system business processes, and U′ = {A′, B′, C′, D′, E′, F′} is the set of information flows among the processes, which in turn are typed by concepts of concept model − a set U = (A, B, C, D, E, F, G) (not shown in Fig. 6.3). Transformation of process model into so-called intermediate model is made in a direct way of graph transformation [4], where the edges of graph of business processes are transformed into nodes of intermediate model. Finally, the nodes of graph of business processes are transformed into the edges of intermediate

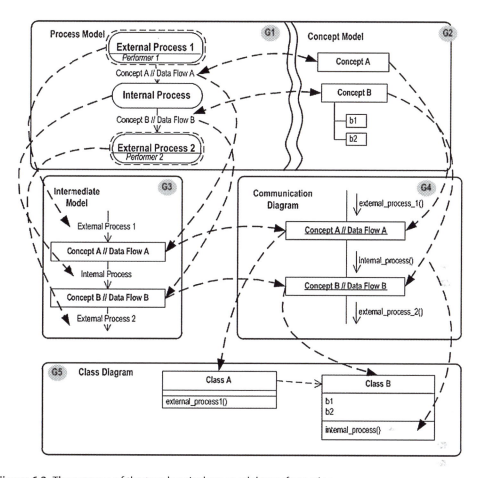

Figure 6.2 The essence of the two-hemisphere model transformation.

model. Then, an intermediate model is redrawn according to notational conventions of UML communication diagram (graph G4 in Fig. 6.2 and 6.3), using a set of concepts, defined in concept model (model G2 in Fig. 6.2). Graph G3 (P, U) corresponds to UML communication diagram, as discussed above. There is a set of nodes U = {A, B, C, D, E, F, G}, which is based on the same set of elements of concept model and represents objects, received from the set U', and set of arcs P = {P1, P2, P3, P4, P5}, which represent interactions between objects in a form of methods to perform during object interaction. During the transition from G1 to G4 through intermediate model each process and each data flow is transferred into corresponding elements of a UML communication diagram. Then the transformations can be applied for generation of the UML class diagram, namely, to generate class names, attributes, methods and several kinds of class relationships [41].

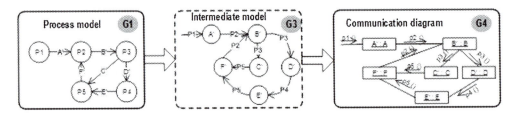

Figure 6.3 Graph transformations from graph of process model through intermediate model into UML communication diagram.

The choice and design of the two interrelated models is not only based on the analogy with the human brain, but also due to the fact that information shown in these diagrams helps to describe the system from the two different points of view that capture the most relevant aspects in system development. Business process modelling, as Polak [42] mentioned, developed as a result of solutions made by Management Science and Computer Science in the 1970s, and keeps showing its importance in supporting process modelling and analysis. The importance of business process modelling is confirmed by Harmon and Wolf [43] researches about the usefulness and usability of these processes. Results confirm that management of business processes is valuable for companies and that firms progressively adopt existing business process modelling notations and methodologies.

Consequently, the two-hemisphere model benefits from the incorporation of well-known business process tools and narrows, at the same time, the gap between existing and needed internal knowledge and procedures. As the two-hemisphere model serves as a bridge between problem domain and software design phase, business model is understandable to both sides — business people and developers. The inclusion of a conceptual model in the approach is motivated by the principles of the object–oriented paradigm and general context of data analysis. In many widely accepted software design approaches, in the first phases of the development cycle a data dictionary is created or an analogous document defines a shared agreement about terminology used in software development and documentation. Ref. [44] describes conceptual modelling as the basis of software development, absolutely needed to produce a quality design. Conceptual models are high-level software description, which contains concepts and relationships among those concepts. Any kind of things, events and living beings that are important to a given problem domain can be considered as concepts. Concepts are described with attributes, but methods indicate actions specific to these concepts. An early and consolidate example of conceptual models is Peter Chen's [40] Entity-Relationship (ER) diagrams, used in database design [45], and, later, in software system design. Consequently, a natural choice for the other part of the two-hemisphere model is a conceptual model, consisting of concepts and related attributes, with a model notation similar to ER diagrams. Currently, the two-hemisphere model gives an ability to generate UML use case, sequence, activity, state and class diagram [4,41,46], some of them are supported by BrainTool [47], the rest is under development. Moreover, the problem on generation of program code has been continuously investigating in the last decade, where the results are published in [48] and [49].

6.5. The two-hemisphere model-driven approach for solving composition problems

The strategy of the two-hemisphere model-driven approach supports model transformation from problem domain models into program components, where problem domain models reflect two fundamental things: system functioning mechanisms (processes) and structure (concepts and their relations). Several two-hemisphere models presenting different aspects of CPS and its components structure and behaviour have been marked as input with mapping rules, the class diagram and transformation trace has been received on output (see Fig. 6.4). A transformation trace shows how an element of the two-hemisphere model is transformed into the corresponding element of the class diagram, and which parts of the mapping are used for transformation of every part of the two hemisphere model [4].

The model decomposition into small components and their composition as an integrated system is a new research topic for the two-hemisphere model-driven approach, originally introduced in [50]. The work is under ongoing development and evolution, consequently, there is still no mature foundation to date for this. Our goal through the research on the composition of CPS is to study existing models of composition approaches by analysing and identifying: a) what are the elements involved in the composition process, and b) how the model composition is made in these approaches.

The ultimate goal is to arrive at an understanding of what is done for model composition in these approaches [51]. According to them, it can be said that the composition model is a process that takes two or more input models and integrates them through an operation and composition to produce a composite output model. However, this scheme is very abstract. No assumptions about the input models, output, or on the compositing operation are expressed. In practice, each approach must specify these assumptions for its work context. These also include the differences to classify approaches:

- composition mechanism: melting, replacing the union, weaving, etc.;
- element composition: what are the additional elements involved in the composition; there are two classification axes: the type and formality of these;
- composition language: The composition of elements needs formalisms to express them.

These formalisms are very diverse because each approach has its own elements of composition. They can be a weaving language, a metamodel of composition rules, a UML profile for model composition, etc. Despite their diversity, they can usually identify a composition formalism on two points: the composition that provides abstractions and the composition that provides scalability.

To synthesise, we can define the composition as a model management operation, which generates a single model by the combination of the contents of at least two models from the following ones:

- syntactic level: expression model compound from input models;
- semantics level: assigning a semantic model compound, depending on the semantics of the associated source models;
- methodical level: using the model compound, derived from the composition process in a software development process.

Therefore, the composition process cannot be considered as an atomic operation. Before triggering the composition process itself it is necessary to identify the links between the elements composing; hence the emergence of the pre-match phase followed by a composition operation that aims at the creation of the model "global" by combining elements using input patterns of relationships defined in the matching pattern.

So, considering all the requirements and specifics mentioned above, it is clear that making a survey on composition techniques and identify their gaps seems an interesting path to build a new models composition operation based on two hemisphere model approach. In other words, we suggest using this taxonomy to create a novel composer framework to resolve composition conflicts for a given problem. Furthermore, we are also studying a way to take into account the semantic properties of models. If we take the example of two operations in two models that appear with the same signature (name, type, parameters), in order to remedy this problem, we must either include a step of reconciliation between integrating the separate designs or strengthen semantics associated with the input metamodel, so that we can implement finer comparison strategies that address the behaviours described by the methods. So far, we studied in recent work the composition of a structural diagram based on the two-hemisphere model-driven approach, in this work we are going to use this method in studying the behavioural side by merging two activity diagrams in order to highlight behavioural modelling.

Behavioural modelling is also important in the design process of a complex system, especially in the context of Model Driven Architecture (MDA), where the goal is to achieve automation of the post-design phases (coding, integration, validation, etc.). Indeed, such automation requires a design model as complete as possible. Behavioural modelling in UML can be done at several levels of abstraction, starting from the overall models such as the interaction and activity models that represent the interactions and the sequence of activities between the different objects or components of the UML system, and going as far as the fine description of the behaviour of objects or components by state machines. The overall models, such as sequence diagrams, allow by definition the description of a behaviour from a point of view or a combination of several points of view.

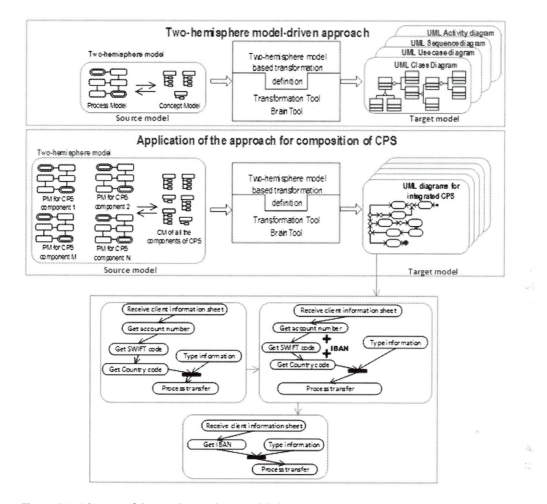

Figure 6.4 Adoption of the two-hemisphere model-driven approach to composition of CSP.

In this part, automated merge are proposed for restructuring UML activity models (as a behaviour model) together based on the two–hemisphere model-driven approach. Modelling a system amounts to determining its static structure and behaviour based on the following three standard perspectives [52]:

- modelling the static structure: this perspective declares the object families and the relationships that connect them. It shows the structural elements such as classes, associations, interfaces, and attributes;
- modelling inter-object behaviours: This overview is useful in the specification of needs in the initial phases of analysis–design. It shows the interactions between the potential objects of the system. It gives partial visions of the global system in the

form of scenarios. By their partial and intuitive nature, the scenarios are the most adapted to express the requirements;

• modelling of intra–object behaviour: this perspective focuses on the representation of the life cycle common to objects of a same class. It describes the complete behaviour of an object by showing the different states and transitions possible at runtime.

UML is intended to describe and represent needs, specify and document systems, and design solutions. It unifies both notations and object-oriented concepts. In its current version, UML 2.5.1 is structured around thirteen types of diagrams, each of them dedicated to the representation of particular concepts of a software system: these types of diagrams are divided into two groups, structural diagrams and behavioural diagrams. In the task of composition of CPS we will focus on behavioural diagrams. UML behavioural diagrams include use case diagrams, activity diagrams, state machine diagrams, sequence diagrams and communication diagrams. When developing a complex system, the construction of a global model that simultaneously takes into account all the needs of the actors and the technical requirements imposed is often impossible. Despite the evolution of analysis/design techniques in the field of software engineering, the construction of computer systems remains a difficult task. In reality, several partial models are developed separately and coexist with the associated risks of inconsistency, i.e. the global model must be frequently challenged when needs change. The object–oriented approach and related concepts (encapsulation, inheritance, and polymorphism) have been an important advance for the design of software systems by bringing in particular modularity and reuse. But this approach has limitations when it comes to dealing with complex systems, multidimensional (i.e. multilevel systems: functional, business, technological, etc.) and highly parallel systems. To control this complexity, we use more and more multimodel modelling approaches.

Activity diagrams can sometimes describe two actions or more representing the same notion: an example might be an activity that takes the account number of a client and then an activity that takes the swift code and another activity which takes the country code. Usually there is a big need to combine the three activities in order to get the IBAN code: in this case they can be merged into one activity which combines the full information (IBAN) of the client (as showed in Fig. 6.4). This presents an advantage for the user as it may be subject to errors in doing this operation manually especially in the bank environment. By moving all edges from one activity to another, and then deleting one of them, we should merge several activities into one. In our example, we will apply the merge method in order to merge three activities into one, our activities are linked together by an edge. The source activity of the connecting edge will be merged into the next activity and then the result will be merged to another activity by doing a subsequent remove for the merged activity. The merge of two activities needs a precondition to be executed, the merged activities must be linked by a flow edge in the right scheduling. The steps for achieving this merge are:

- move the source nodes from the first activity to be removed to the second activity that will be kept potentially;
- move the target nodes from the first activity to be removed to the second action that will be kept potentially;
- move the source nodes from the second activity kept potentially and to be removed to the third activity that will be kept definitely;
- move the target nodes from the second activity kept potentially to be removed to the third activity that will be kept definitely;
- add name of the second activity to be removed to the merged activity (third action);
- remove the control flow node that connects the activities to be merged;
- remove subsequently the first and the second activity.

6.6. Conclusions

CPS compositional approaches are a promising and viable strategy for a consistent and manageable definition of large complex CPS systems [1]. In this chapter we presented a possible model composition technique for CPS based on the two-hemisphere model-driven approach introduced in [3]. The central idea of the two-hemisphere model-driven approach is to apply several transformations for composition of the complex system from small parts, where every of them presented as the source model defined in terms of a business process model, associated with a concept model, and the target model defined in terms of UML diagrams, which are generated for the whole system. When the models are small enough and developed by a single or a couple of designers, they can be composed manually. However, in the case of a cyber-physical system, the models are too large to be composed manually and it's necessary to develop an automatic composition method to ensure that all the elements in the model are handled.

As far as the two-hemisphere model-driven approach can be applied to provide a model composition paradigm we assumed to use it to support composition automation since the very first phases of the design process, directly working on the class diagrams of the system. This approach proved to be suitable to correctly capture and represent knowledge as it is needed and desired by the business users, but as it is profitable and useful to system analysts and designers at the same time: it supports definition of architectural designs, assignment of responsibilities, representation of structural relationships and dependencies and component-oriented issues, and allows the definition of a detailed system architecture by means of automatic transformations applied to component descriptions. Consequently, we expect that this allows one to master the complexity of CPS by leveraging automation that preserves and comply by design all system requirements. The conceptual framework of the two-hemisphere approach will be extended towards fine-grain component-oriented specifications management and will be implemented into a proper ATL based software framework, aiming to open source public distribution of the software tools behind the approach.

By applying this method, it is possible to automate the process of UML diagram development for a CPS starting from a two-hemisphere model-based description of the system including needed structural and behavioural information in a representation that is easy to understand for both business users and system analyst. Using the two-hemisphere model-driven approach native mechanisms to formally represent shared responsibilities among object classes and inter-class relationships definition, it is possible to define, for a given CPS that needs to be designed or analysed:

1) a description of its organisation in terms of its components (as it is possible by means of classes in object-oriented representations);

2) how responsibilities should be shared between the components, i.e., in terms of processes to be jointly and co-ordinately enacted by given components;

3) the structural relationships and interdependencies between the components of the CPS, needed for the implementation process.

Future research will explore the nature of the transformation process itself, to investigate the reversibility of the design process to better understand its definition and to extend the framework to automatically include existing component descriptions based on other metamodels or heterogeneous in other description aspects.

Acknowledgements

This publication was developed in the framework of COST Action IC1404 – Multi-Paradigm Modelling for Cyber-Physical Systems, and with minor revisions is published by Departamentos Lenguajes y Ciencias de la Computación, Universidad de Málaga in Proceedings of the 4th Workshop of the MPM4CPS COST Action as Technical Report No. ITI16/01.

References

[1] K. Babris, O. Nikiforova, U. Sukovskis, Brief overview of modelling methods, life-cycle and application domains of cyber-physical systems, Applied Computer Systems 24 (1) (2019) 5–12.

[2] O. Nikiforova, N. El Marzouki, K. Gusarovs, H. Vangheluwe, T. Bures, R. Al-Ali, M. Iacono, P. Orue Esquivel, F. Leon, The two-hemisphere modelling approach to the composition of cyber-physical systems, in: Proceedings of International Conference on Software Technologies (ICSOFT 2017), Madrid, Spain, 24–26 July, 2017, SCITEPRESS Digital Library, 2017, pp. 286–293, https://doi.org/10.5220/0006424902860293.

[3] O. Nikiforova, M. Kirikova, Two-hemisphere driven approach: engineering based software development, in: Advanced Information Systems Engineering, Proceedings of the 16th International Conference CAiSE 2004, Springer Verlag, Berlin, Heidelberg, 2004.

[4] O. Nikiforova, Two hemisphere model driven approach for generation of UML class diagram in the context of MDA, in: Z. Huzar, L. Madeyski (Eds.), e-Informatica Software Engineering Journal 3 (1) (2009) 59–72.

[5] S.K. Khaitan, J. Mccalley, Design techniques and applications of cyber physical systems: a survey, IEEE Systems Journal 9 (2) (2014) 1–16.

[6] F. Hu, Cyber-Physical Systems: Integrated Computing and Engineering Design, CRC Press, New York, 2018, 398pp.

[7] D. Patterson, J. Hennessy, Computer Organization and Design: The Hardware Software Interface, 5th edition, Morgan Kaufmann, 2013, 793pp.

[8] E.A. Lee, The past, present and future of cyber-physical systems: a focus on models, Sensors 15 (3) (2015) 4837–4869.

[9] E.A. Lee, S.A. Seshia, Introduction to Embedded Systems - A Cyber-Physical Systems Approach, LeeSeshia.org, 2011.

[10] E. Sultanovs, A. Romanovs, Centralized healthcare cyber-physical systems data analysis module development, in: Proceedings of the 2016 IEEE 4th Workshop on Advances in Information, Electronic and Electrical Engineering, Lithuania, Vilnius, 10–12 November 2016, 2016.

[11] A. Romanovs, I. Pichkalov, E. Sabanovic, J. Skirelis, Industry 4.0: methodologies, tools and applications, in: Proceedings of the Open International Conference on Electrical, Electronic and Information Sciences eStream 2019, Lithuania, Vilnius, 25–25 April, 2019.

[12] R. Buyya, J. Broberg, A. Goscinski, Cloud Computing: Principles and Paradigms, John Wiley & Sons, 2010, 637pp.

[13] S. Kim, S. Park, CPS (Cyber Physical System) based manufacturing system optimization, Procedia Computer Science 122 (2017) 518–524, https://doi.org/10.1016/j.procs.2017.11.401.

[14] A. Teilans, A. Romanovs, J. Merkurjevs, P. Dorogovs, A. Kleins, S. Potryasaev, Assessment of cyber physical system risks with domain specific modelling and simulation, SPIIRAS Proceedings 4 (59) (2018) 115–139.

[15] A. Romanovs, Security in the era of Industry 4.0, in: 2017 Open Conference of Electrical, Electronic and Information Sciences (eStream), 2017, p. 1.

[16] O. Hancu, V. Maties, R. Balan, S. Stan, Mechatronic approach for design and control of a hydraulic 3-DOF parallel robot, in: The 18th International DAAAM Symposium, "Intelligent Manufacturing & Automation: Focus on Creativity, Responsibility and Ethics of Engineers", 2007.

[17] S.C. Suh, J.N. Carbone, A.E. Eroglu, Applied Cyber-Physical Systems, Springer, 2014.

[18] C.-R. Rad, O. Hancu, I.-A. Takacs, G. Olteanu, Smart monitoring of potato crop: a cyber-physical system architecture model in the field of precision agriculture, in: Proceedings of the Conference Agriculture for Life, Life for Agriculture, 2015.

[19] S.K. Khaitan, J.D. McCalley, Design techniques and applications of cyber physical systems: a survey, IEEE Systems Journal 9 (2) (2015).

[20] F.-J. Wu, Y.-F. Kao, Y.-C. Tseng, From wireless sensor networks towards cyber physical systems, Pervasive and Mobile Computing 7 (4) (2011).

[21] T. Bureš, I. Gerostathopoulos, P. Hnětynka, J. Keznikl, M. Kit, F. Plášil, DEECo - an ensemble-based component system, in: Proceedings of CBSE 2013, 2013.

[22] R. Hennicker, A. Klarl, Foundations for Ensemble Modeling – The Helena Approach, Specification, Algebra, and Software, Lecture Notes in Computer Science, vol. 8373, 2014.

[23] L.P. Carloni, F. De Bernardinis, C. Pinello, A.L. Sangiovanni-Vincentelli, M. Sgroi, Platform-based design for embedded systems, http://www.cs.columbia.edu/~luca/research/pbdes.pdf, 2005.

[24] M. Gribaudo, M. Iacono, An introduction to multiformalism modelling, in: Theory and Applications of Multi-Formalism Modelling, IGI-Global, 2014.

[25] H. Vangheluwe, J. De Lara, P.J. Mosterman, An introduction to multi-paradigm modelling and simulation, in: Proceedings of the AIS'2002 Conference (AI, Simulation and Planning in High Autonomy Systems), 2002.

[26] E. Barbierato, M. Gribaudo, M. Iacono, Exploiting multiformalism models for testing and performance evaluation in SIMTHESys, in: Proceedings of 5th International ICST Conference on Performance Evaluation Methodologies and Tools - VALUETOOLS 2011, 2011.

[27] E. Barbierato, M. Gribaudo, M. Iacono, Modeling hybrid systems in SIMTHESys, in: Electronic Notes on Theoretical Computer Science, Elsevier, 2016.

[28] E. Barbierato, G. Dei Rossi, M. Gribaudo, M. Iacono, A. Marin, Exploiting product forms solution techniques in multiformalism modelling, in: Electronic Notes in Theoretical Computer Science, Elsevier, 2013.

[29] J. Keznikl, T. Bureš, F. Plášil, M. Kit, Towards dependable emergent ensembles of components: the DEECo component model, in: Proceedings of WICSA/ECSA 2012, 2012.

[30] S. Nazari, C. Sonntag, S. Engell, A Modelica-based modelling and simulation framework for large-scale cyber-physical systems of systems, IFAC PapersOnLine 48 (1) (2015).

[31] A. Bhave, B.H. Krogh, D. Garlan, B. Schmerl, View consistency in architectures for cyber-physical systems, in: Proceedings of IEEE/ACM International Conference on CPS (ICCPS), IEEE, 2011.

[32] T. Bureš, P. Hnetynka, J. Kofron, R. Al Ali, D. Škoda, Statistical approach to architecture modes in smart cyber physical systems, in: Proceedings of WICSA 2016, IEEE, 2016.

[33] T. Bureš, F. Krijt, F. Plášil, P. Hnětynka, Z. Jiráček, Towards intelligent ensembles, in: Proceedings of the 9th European Conference on Software Architecture Workshops (ECSAW 2015), ACM, 2015.

[34] R. Al Ali, T. Bureš, I. Gerostathopoulos, J. Keznikl, F. Plášil, Architecture adaptation based on belief inaccuracy estimation, in: Proceedings of the 11th Working IEEE/IFIP Conference on Software Architecture (WICSA 2014), 2014.

[35] M.A. Ould, Business Process Management: A Rigorous Approach, British Computer Society, 2005.

[36] A. Badica, C. Badica, F. Leon, I. Buligiu, Modeling and enactment of business agents using Jason, in: Proceedings of the 9th Hellenic Conference on Artificial Intelligence, SETN 2016, 2016.

[37] F. Leon, C. Badica, A comparison between Jason and F# programming languages for the enactment of business agents, in: Proceedings of the International Symposium on Innovations in Intelligent Systems and Applications, 2016.

[38] O. Nikiforova, General framework for object-oriented software development process, in: Scientific Proceedings of Riga Technical University, in: Computer Science, Applied Computer Systems, vol. 13, Riga, 2002, pp. 132–144.

[39] J. Anderson, Cognitive Psychology and Its Implications, W.H. Freeman and Company, New York, 1995.

[40] P. Chen, The entity relationship model—towards a unified view of data, ACM Transactions on Database Systems 1 (1976) 9–36.

[41] K. Gusarovs, O. Nikiforova, Workflow generation from the two-hemisphere model, Applied Computer Systems (ISSN 2255-8683) 22 (2017) 36–46, https://doi.org/10.1515/acss-2017-0016.

[42] P. Polak, BPMN impact on process modelling, in: Proceedings of the 2nd International Business and Systems Conference BSC, 2013.

[43] P. Harmon, C. Wolf, The State of Business Process Management, BPTrends, 2014, http://www.bptrends.com/.

[44] J. Johnason, A. Henderson, Conceptual Models. Core to Good Design, 1st edition, Morgan & Claypool Publishers, 2011.

[45] W. Hesse, Ontologies in the software engineering process, in: Proceedings of the 12th International Workshop on Exploring Modelling Methods for Systems Analysis and Design (EMMSAD-2007), 2007.

[46] O. Nikiforova, K. Gusarovs, A. Ressin, An approach to generation of the UML sequence diagram from the two-hemisphere model, in: H. Mannaert, et al. (Eds.), Proceedings of the 11th International Conference on Software Engineering Advances, ICSEA 2016, August 21–25, 2016, Rome, Italy. ©IARIA, pp. 142–148, available at http://www.thinkmind.org/.

[47] O. Nikiforova, K. Gusarovs, Comparison of BrainTool to other UML modeling and model transformation tools, in: AIP Conference Proceedings, International Conference on Numerical Analysis and Applied Mathematics 2016, ICNAAM 2016; 6th Symposium on Computer Languages - SCLIT 2016, Rhodes, Greece, 19–25 September 2016, 2017, pp. 19–25.

[48] K. Gusarovs, O. Nikiforova, A. Giurca, Simplified lisp code generation from the two-hemisphere model, Procedia Computer Science (ISSN 1877-0509) 104 (2017) 329–337, https://doi.org/10.1016/j.procs.2017.01.142.

[49] O. Nikiforova, K. Gusarovs, Anemic domain model vs rich domain model to improve the two-hemisphere model-driven approach, Applied Computer Systems (ISSN 2255-8683) 25 (1) (2020) 51–56, https://doi.org/10.2478/acss-2020-0006.

[50] N. El Marzouki, O. Nikiforova, Y. Lakhrissi, M. El Mohajir, Enhancing conflict resolution mechanism for automatic model composition, in: J. Grundspenkis, et al. (Eds.), Scientific Journal of Riga Technical University: Applied Computer Systems 19 (2016) 44–52.

[51] L. Cavallaro, E. Di Nitto, C.A. Furia, M. Pradella, A tile-based approach for self-assembling service compositions, in: Engineering of Complex Computer Systems (ICECCS), 2010.

[52] J. Krogstie, A. Sølvberg, Information Systems Engineering – Conceptual Modeling in a Quality Perspective, Kompendiumforlaget, 2003.

CHAPTER 7

Multi-paradigm modelling and co-simulation in prototyping a cyber-physical production system

Mihai Neghină[a], **Constantin Bălă Zamfirescu**[a], **Peter Gorm Larsen**[b] **and Ken Pierce**[c]

[a]Lucian Blaga University, Sibiu, Romania
[b]Aarhus University, Aarhus, Denmark
[c]School of Computing, Newcastle University, Newcastle upon Tyne, United Kingdom

7.1. Introduction

In the development of truly complex Cyber-Physical Systems (CPSs), a model-based approach can be an efficient way to master system complexity through an iterative and incremental development. Such systems are often made in an ad hoc fashion by combining different CPSs without explicit and comprehensive coordination among the design teams. The expertise and engineering background for such design teams are often limited to the domain of their respective subsystem. These domains include their own focus, terminology and modelling approaches. Engineers of physical systems often use continuous-time (CT) formalisms, realised as differential equations, to produce high-fidelity simulation of physical phenomena. Meanwhile, software engineers tend to adopt discrete-event (DE) formalisms, focusing on the logical behaviours of control systems. To a large extent, these CPSs emerge and evolve through iterative and incremental developments, from digital model through multiple, costly physical prototypes where teams are unable to truly collaborate and design faults found at a late stage are extremely costly.

In this chapter we illustrate how a co-simulation technology [1] can be used to gradually increase the detail in a collaborative model (co-model) following a "discrete event first" (DE-first) methodology [2]. In this approach, initial abstract models are produced using a discrete event (DE) formalism (in this case VDM) to identify the proper communication interfaces and interaction protocols among different models. These are gradually replaced by more detailed models using appropriate technologies, for example continuous-time (CT) models of physical phenomena.

The case study deals with the virtual design and validation of a CPS-based manufacturing system for assembling USB sticks. It is a representative example of part of a distributed and heterogeneous system in which products, manufacturing resources, orders and infrastructure are all cyber–physical. In this setting, several features (such as asynchronous communication, messages flow, autonomy, self-adaptation, etc.) would be investigated at design time, for example using a collaborative modelling approach. Consequently, the case study offers a balance between being sufficiently simple to be easily followed as a production line example, including generating a tangible output, and at the same time being sufficiently general to allow the study of the co-simulation complexity [3]. Furthermore, by choosing the production of USB-OTG[1] sticks shown in Fig. 7.1, the example opens the possibility of extending the purpose of the study to interactions between generated hardware and generated software solutions in the production line.

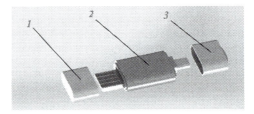

Figure 7.1 Example of an USB-OTG stick consisting of three component parts: 1) left cap; 2) middle (body) of the stick; 3) right cap.

7.2. Case study description

The case study was developed as part of Integrated Product-Production co-simulation for Cyber-Physical Production System (IPP4CPPS) innovation experiment [4]. The ambition of iPP4CPPS project was to contribute to the advancement of engineering methods and tools employed in manufacturing CPS-based production systems by:

- Demonstrating the proper methodological steps for achieving a working heterogeneous co-simulation (with units modelled in various dedicated tools) of a relatively complex system that requires diverse and multidisciplinary teamwork.
- Evaluating the maturity level of current CPS-based technologies for future implementation into CPS-based production systems;
- Extending the libraries and functionalities of the employed tools to cope with the real industrial needs.

This section provides a brief description of the case study production line for manufacturing of USB-OTG sticks as a tractable but representative production line example. This production line has the classical characteristics of a smart product, as defined by Mühlhäuser [5]:

---

[1] A USB flash drive or stick with On-The-Go (OTG) technology is equipped with a standard USB connector on one end and a micro-USB connector on the other end, allowing direct connection to devices such as smartphones.

- Situated: recognition of situational context, in terms of order identification, availability of parts and slots, awareness of perturbations (e.g. vibrations) and malfunctions, etc.
- Personalised: personalisation of USB-OTG sticks according to orders, as well as the capability of handling cancellations and order modifications during the production phase.
- Adaptive: adaptation of the line to the customer orders, for instance according to order urgency and the level of perturbations (vibrations).
- Pro-active: anticipation of the intentions of the production line owner by restricting functionality in certain conditions in order to minimise the risk of malfunctioning or extending the testing in uncertain conditions of luminosity.
- Business-awareness: energy-efficient behaviour unless receiving special urgency requests by the customers.
- Network capable: although not tested in this experiment, each production unit has intrinsic communication capabilities with external products (including similar production lines).

To capture the value-adding processes in Industry 4.0 [6–9], the case study includes distinct subsystems to reflect order-placing users, as well as the required infrastructure that enables the other CPSs to function properly.

The subsystems identified as necessary for the case study and shown in Fig. 7.2 are:
- The Human–Machine Interface (HMI), which is handling incoming orders, being responsible for interpreting and transmitting them correctly to the Part Tracker.
- The Part Tracker, which is the infrastructure unit capable of communicating with the HMI, capable of relaying the order information to the production system and of gathering data on the status of any order received.
- The Warehouse, which assembles the sticks from the stored parts.
- The Robotic Arm capable of moving parts or assembled sticks.
- The Wagons, which are the transportation units between subsystems of the production line.
- The Test Station, which is the processing station for checking the conformity to order requirements.

An additional visualisation module for the system units has been created in Unity. Fig. 7.2 shows the 3D rendering of the production line layout and the small demonstration stand with the physical units realised during the project.

The communication patterns between units, represented in Fig. 7.3, contain both simple and composed messages. Simple messages are used on dedicated lines for straightforwardly transmitting the state of the subsystem or a certain value. Examples include the requests from the Part Tracker to the Wagons to assume a certain speed or the feedback of the Wagon positions to the Part Tracker.

The purpose of the composed messages is twofold: to ensure that certain bits of information arrive simultaneously (as opposed to them coming on different message lines

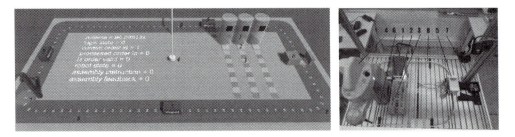

Figure 7.2 Layout of the Unity rendering and the physical demo stand, containing: 1) the Warehouse stacks; 2) the Warehouse assembly box; 3) the Warehouse memory boxes; 4) the Robotic Arm; 5) Wagons on the track; 6) the Loading Station; 7) the Test Station; 8) the circular track for the Wagons.

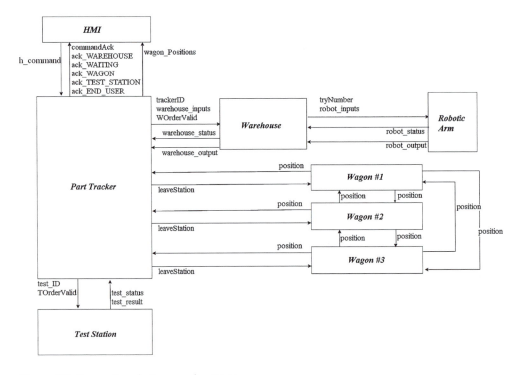

Figure 7.3 Connections between subsystems.

that may become unsynchronised or for which further synchronisation logic might be needed) and to account for the possibility of coded messages (that might be interesting in applications with significant noise, where error correcting codes might become useful). The order requests from the HMI to the Part Tracker or their acknowledgement contain multiple pieces of information in each (order ID, urgency, user choices, and so on).

7.3. Techniques

This section introduces the main technologies used for building the initial models, including the INTO-CPS co-simulation technology and the way of generating Discrete-Event First (DE-first) models in Overture [10]. This section also presents shortcomings in directly applying the workflow of INTO-CPS and the motivation for an alternative approach.

7.3.1 The INTO-CPS technology

The INTO-CPS co-simulation technology is a collection of tools and methods that have been linked to form a tool chain for model-based design of CPSs. Rather than suppressing the diversity of the software, control and mechatronics formalisms necessary in the design of CPSs by requiring all disciplines adopt the same general-purpose notation, INTO-CPS embraces this diversity by integrating them at a semantic level [11–15], allowing engineers to collaborate using familiar modelling techniques and methods. The overall workflow and services from the tool chain used in this project are illustrated in Fig. 7.4.

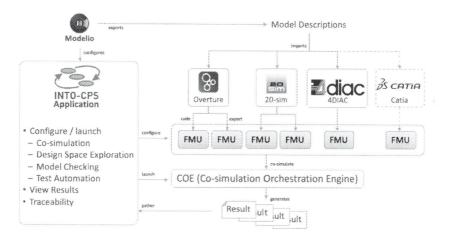

Figure 7.4 The INTO-CPS tool chain used in this project.

The INTO-CPS tool chain is centred around co-simulation of heterogeneous models using the Functional Mock-up Interface (FMI) standard [16] which allows models from different tools and formalisms to be packaged as Functional Mock-up Units (FMUs). Such models can be combined and analysed through co-simulation, with each FMU acting as an independent simulation unit and having a model description that includes its interface. FMUs can also be provided as a black-boxes to protect intellectual property (IP) contained in the details of the model.

INTO-CPS includes a co-simulation engine, called Maestro [17], that fully implements version 2.0 of the FMI standard and has been successfully tested with over 30 tools [16]. Around this co-simulation core, the INTO-CPS tool chain links additional tools to support model-based design throughout development. A Systems Modelling Language (SysML) profile is provided and supported by the Modelio tool [18], allowing FMU model descriptions to be captured and linked to requirements. The model descriptions can capture both physical and cyber parts of the system, enabling both Hardware-in-the-Loop (HiL) and Software-in-the-Loop (SiL) simulation, and can be exported for use in modelling tools [6]. The profile also allows these descriptions of FMUs to configure co-simulations and other forms of analysis supported by INTO-CPS.

Specific support for importing model descriptions and producing skeleton models is provided by tools in the INTO-CPS tool chain: Overture [10], supporting DE modelling; 20-sim [19] and OpenModelica [20], both supporting CT modelling. These tools also guarantee export of FMUs, which can then be used in a co-simulation with Maestro [17]. FMU export is included in an increasing number of industry tools. During the heterogeneous co-simulation in the later stages of our experiment we used 4DIAC [21], an open-source tool for development of industrial control systems, and CATIA [22], an industry-standard software suite for computer-aided design and manufacturing. Both are shown in their respective positions within the INTO-CPS tool chain in Fig. 7.4.

7.3.2 Initial models

The approach of directly generating FMI model descriptions from SysML profiles and importing them into dedicated modelling tools, with the expectation that generated FMUs would seamlessly combine in the co-simulation, can be prone to failure. While allowing different teams to work on constituent models separately, it requires FMUs from all teams to be available before integration testing through co-simulation. Any delay in the generation of any component leads to further delays in the co-simulation and to the late discovery of problems. Similarly, if there is a single team producing all FMUs sequentially, in their full complexity, the co-simulation can only begin at the end of modelling.

A potential strategy to mitigate these risks is to have each team produce quick, initial versions of FMUs as soon as they can and perform integration testing with these models. The initial models can then be updated in an iterative manner towards more detailed models, with each team able to move at its own pace, and with previous versions providing fall-backs in the case of problems and baselines for regression testing. This approach might be more difficult in some modelling paradigms, however, where quick and simple models might not function sufficiently well for testing.

The DE-first approach adopted for the IPP4CPPS project follows the strategy of producing initial FMUs and replacing them with more detailed models as they become available. Rather than using each individual formalism, a single DE formalism such as Overture [10] can be used for all initial models, thus quickly generating the abstract model of the whole system. Sketching out the behaviour of constituent models allows the early testing of assumptions at the beginning of the process. A DE formalism should be selected because these are designed to capture abstract and logical behaviours, often described in terms of interfaces, and therefore are well-suited to this task [23].

7.3.3 Discrete-event first with VDM-RT/Overture

Applying a DE-first approach to an FMI setting using Overture [10] follows the principles of the Vienna Development Method (VDM) [24,25], a set of modelling techniques successfully applied in both research and industrial application. The initial VDM Real-Time (VDM-RT) [7] project contains a class for each FMU, with port objects corresponding to the interface given in the model description, a main (system) class that instantiates appropriate port objects and instances of each FMU class to which it passes the ports and a world class that provides a method as an entry point for simulation by starting the threads of the FMU objects and blocks until simulation is complete.

Fig. 7.5 provides class and object diagrams and shows such a set-up using two constituent models, FMU1 and FMU2. Such a model can be simulated within Overture to see how the FMUs behave and interact. Once sufficient confidence in these initial models is gained, they can be exported individually as FMUs and integrated in a co-simulation.

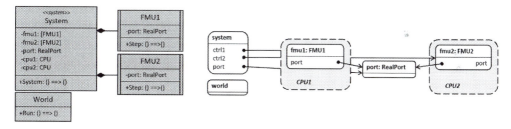

Figure 7.5 Class diagram showing two simplified FMU classes created within a single VDM-RT project, and an object diagram showing them being instantiated as a test.

The Overture FMI plug-in can then be used to export an FMU from each individual project unit, these can then be combined in a co-simulation. These FMUs can be revised if problems are found, then replaced with higher-fidelity models. The models could be retained for later use and as a fall-back in case of future problems in integration.

7.4. Methodology

There are two distinct phases of development: the digital model (Table 7.1) and the construction of the prototype and deployment (Table 7.2) [2,3]. For the first phase, the agent-oriented approach was best suited to provide the most adequate abstractions to design the conceptual model of the prototype by identifying its main subsystem types (i.e. production machines, order, and factory infrastructure) and define the interaction protocols among these subsystems. These types are well-established in agent-based manufacturing control system [26], and are now part of the more complex and abstract Reference Architecture Model for Industry 4.0 [27].

Table 7.1 The digital model development phase.

Stages	Goals
The requirements model: a detailed preliminary mechanical model of the production demo (domain description).	• to identify the compositional structure of the targeted co-simulation and the best-suited model/simulation tool for each component from the production demo; • to facilitate a shared understanding among the specialised teams engaged in implementing the specific simulations.
The homogeneous co-simulation model: the high-level abstraction for the behaviour of each simulation, and the interactions among the constituent simulations.	• to validate the interaction protocols among the constituent simulations; • to have an early working co-simulation where the specific simulations may be gradually added, tested and validated; • to lessen the dependency among the dispersed teams involved in modelling the specific simulations; • to cover the left-over parts of the co-simulation that are not needed to be modelled at a high-level of details (e.g. the Test Station); • to identify conceptual (system-level) parameters that can be used at a later stage as stable constraints in the design space exploration for fine tuning.

While the first phase was intended to determine how to build the prototype, in the second phase the prototype was implemented in its final form. Therefore, each subsystem was developed in a specific language and tool, suitable to the domain and expertise of the team, by following the component-based approach to reach its concrete realisation. Table 7.3 shows the correspondence between the subsystems and the adequately suited tool for implementing a complete simulation, as well as the deployment devices considered.

7.5. Modelling of the subsystems

The VDM-RT models created during the homogeneous co-simulation phase are not meant to be accurate in the physical (mechanical/electrical) implementation sense. The VDM-RT models do not need to have complete functionality for the homogeneous

Table 7.2 The construction and deployment development phase.

Stages	Goals
The heterogeneous co-simulation model: the detailed model of each simulation. It includes both continuous-time (CT) and discrete-event (DE) models in various simulation tools.	• to simulate, test and validate from a holistic perspective and with an increased level of accuracy an entire system; • to generate code from the specialised simulation tools of the different subsystems for specific hardware realisations.
The deployment model: the units modelled and tested by the heterogeneous co-simulation have been deployed in the demo stand for fine tuning under real-life conditions.	• to extend the libraries (e.g. 20-sim and 4DIAC with specific sensors and communication protocols) and functionalities (e.g. INTO-CPS, Overture with visualisation and code-generation capabilities) of the employed tools to cope with the real industrial needs.

Table 7.3 The simulation technology and deployment infrastructure.

Type	Unit	Simulation Technology	Deployment Device
Orders	HMI	4DIAC + MQTT	Smartphones / tablets
Infrastructure	Part Tracker	Overture (VDM-RT)	NVIDIA Tegra Jetson
Production	Warehouse	20-sim	Raspberry Pi + Stäubli
Production	Wagons	4DIAC	Raspberry Pi + sensors
Production	Test Station	4DIAC	Camera + Raspberry Pi
Overview	Unity	Unity animation	PC

co-simulation, only the bare minimum from which the communication lines between units and the system-level parameters can be validated.

The incompleteness of the VDM-RT models is related to details of the inner workings of the components, not necessarily respecting all the constraints of reality. One type of incompleteness relates to using random variables instead of true computations of inner workings. The Test Station is an illustrative example, as its initial model simply generated a random conclusion whether the stick had been assembled using the requested colour scheme or not. The initial model of the Warehouse generates colour components randomly, as if the stacks of raw parts were infinite and unpredictable, neither of which is true for the physical unit.

The initial model of Warehouse also exhibits another type of incompleteness: the breaking of continuity. The physical Warehouse sequentially drops coloured components from stacks into the assembly box and instructs the Robotic Arm to remove the parts if they do not match the requested colour. If an order is cancelled during assembly, all components would need to be removed by the Robotic Arm, so that the assembly box is cleared for the next order. However, the initial model of the Warehouse immediately starts generating new colours for the next order without instructing the Robotic Arm to remove existing parts.

Such aspects of the functionality are minor details with respect to the DE modelling used for the abstract validation. The internal states, however, are all well-established, along with the communication patterns and lines between modules, such that the behaviour of the refined modules does not diverge substantially from the behaviour of the abstract models. Once established, the communication lines and the types of data they carry become hard constraints of the simulation that cannot be easily changed, but new lines of communication could be added if necessary. For instance, new communication lines have been added later in the development of the project, for transmitting (to the Part Tracker) the level of vibration recorded by each subsystem.

Besides validating the interaction protocols and decomposing the project into units which could then be worked on separately, another advantage of having only VDM-RT models as the first stage of development is the possibility of determining conceptual (system-level) parameters before the design space exploration at the heterogeneous co-simulation stage. The identified parameters of the production line can be grouped into two categories:

- *Conceptual (system-level) parameters*, which are independent of (or do not pose any significant problems for) the physical realisation, such as the number of colour choices available to the customers, the number of memory box slots in the Warehouse or the carrying capacity of the Wagons. These parameters are suited for analysis during the homogeneous co-simulation stage.

- *Physical (subsystem-level) parameters*, which define the physical properties of the components inside subsystems, such as the parameters of the pneumatic piston used in the assembly tray of the Warehouse, mechanical and electrical parameters for the Robotic Arm and Wagons, sensitivity of the sensors used in various parts of the production line, etc. These parameters are suited for analysis during the heterogeneous co-simulation stage.

The conceptual parameters can be determined sufficiently accurately even with the incomplete simulation model. The number of colour choices available to the customers may influence the stocks of parts required for a good functioning of the assembly line but is not influencing the physical design in any way. The number of memory box slots and the carrying capacity of the Wagons influence the design of those units, but without increasing the complexity of the mechanical design. Adding a few slots or a second indentation in the upper surface of the Wagon (for a second USB-OTG stick), although becoming hard constraints for the heterogeneous simulation, poses no additional mechanical difficulty compared to not adding them and therefore does not impact the future stages of development.

With these parameters determined beforehand, the design space exploration at the heterogeneous stage is constrained and focused on identifying the physical properties of the components that go into the mechanical, electrical and other designs of the subsystems.

7.5.1 Subsystem models

HMI

The HMI unit handles the user interface and communicates only with the Part Tracker.

The project generated two complementary implementations of the HMI unit: the Overture (VDM) implementation and the 4DIAC+MQTT implementation. Although they cannot be used both at the same time, each is useful for a specific type of experiment.

The 4DIAC+MQTT implementation allows manual placing, changing or cancelling of orders at any time through apps for smart devices as shown in Fig. 7.6 and is meant for gathering user heuristics and analysing issues related to the real-time placement of orders.

Figure 7.6 Interaction between external apps and the dedicated HMI.

However, more useful for the analysis of the system is the Overture (VDM) model, which allows an automatic feed of orders through a comma-separated values file (*.csv). This approach is both flexible and powerful. The flexibility stems from the possibility of creating scenarios with various amounts of orders for covering statistical possibilities, while the power comes from the repeatability of experiments. Having the same input *.csv file, the co-simulation can be run with various parameters, but having the same input orders at the same time, thus generating a detailed picture of the behaviour of the system in controlled, repeatable experiments.

Part Tracker

The Part Tracker is the central logical unit that handles orders, as well as gathering and aggregating statistical data. It communicates with all other units except the Robotic Arm. The Part Tracker is the only subsystem modelled in full, being kept as an Overture/VDM-RT model or Overture-generated FMU throughout the project.

Warehouse

The Warehouse is responsible for the storage of components and the assembly of USB-OTG sticks from such parts. It communicates only with the Part Tracker and the Robotic Arm.

As shown in Fig. 7.7, it contains stacks for each type of component, an assembly box for the actual assembly of the items and memory boxes for storing components that do not fit the current order. The memory boxes may also be a source of component parts for new orders, if the requested colour is available.

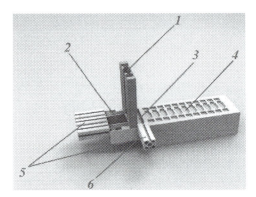

Figure 7.7 Mechanical design of the Warehouse, highlighting: 1) storage stacks; 2) actuators that push available parts from the stack into the assembly tray; 3) the assembly tray; 4) memory boxes with 11 slots per part type; 5) colour detection sensors; 6) pneumatic actuators for the assembly.

For the homogeneous phase analysis, the simulated Warehouse is considered to have an infinite number of parts in the stacks (i.e. there would be no need to re-fill the stacks at any time). Also, the memory boxes would have the same size for all part types (left, middle and right). Upon receiving an assembly order, the Warehouse first looks in the memory boxes for available parts of requested colours. If these do not exist, either because the memory boxes are empty or filled with differently coloured parts, the Warehouse drops the parts from the stacks. However, the parts in the stacks are not arranged in any particular order. Unless it is a lucky hit, a dropped part has the wrong colour and would need to be stored in the memory boxes (if space is available) or thrown for recycling (if memory boxes are full). To keep a reasonable symmetry, all part types have the same range of colours for the users to choose from. Up to eight colours can be selected for the system, but if a colour is available for a component part, it is also available for all other component parts.

Wagons

The purpose of the Wagons is to carry the sticks from the loading station (near the Warehouse) to the Testing Station. The wagons communicate with the Part Tracker and each other. Each Wagon can be certain of its location only at the loading or Test

Station, while in between the position is estimated periodically from previously known position, time and speed.

The system has been restricted to having three Wagons, but each of them has the capacity of carrying one or two assembled sticks. The Wagons cannot overtake each other and their initial model does not cover their malfunction (e.g. fall off the tracks or losing the load). The mechanical design of a double capacity Wagon is shown in Fig. 7.8.

Figure 7.8 Mechanical design of a Wagon containing: 1) electronic board for control; 2) electronic motoreductor; 3) drive wheel; 4) sensor for station detection; 5) driven wheels; 6) motor driver; 7) ultrasonic distance detectors; 8) slots for transporting sticks.

Robotic Arm

The Robotic Arm, shown in Fig. 7.9, moves parts or pieces from one location to another, either because the current available part is not required or because the stick has been assembled and needs to be moved to a loading station. It communicates only with the Warehouse. As with other models, the Overture/VDM-RT model of the Robotic Arm is on purpose incomplete, since the time of any action is not taking into account the physical distance of the endpoints of the action.

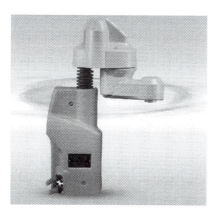

Figure 7.9 Stäubli TS20 robotic arm.

Test Station

The Test Station reads the item characteristics and reports on their conformity to requirement. It communicates only with the Part Tracker, from which it received the requested set of colours for the item and to which it reports the test result. In the VDM-RT model, the testing time is set to a fixed time and the test results are random. However, the rejection rate can be controlled from the model.

7.6. Verification & validation

This section includes experiments performed for each of the development phases, together with the conclusions inferred from them and the advantages of the methodology as observed with the implementation of the USB-OTG production line.

7.6.1 Homogeneous phase experiments

Having a system covering all the important stages of a production line, from orders to delivery, enabled two types of analysis scenarios:

- Analysis of the behaviour of the system when varying parameters external to the system (exogenous parameters)
- Analysis and optimisation of the system when varying parameters internal to the system (endogenous parameters)

System setup

All simulated experiments have been run for 45000 time units, which correspond to 45000 seconds or 12.5 hours for the physical systems (although the simulation itself took about 6 hours for each experiment). The orders can be placed at any time within the initial 12 hours, while the remaining 30 minutes were added to allow for the completion of the late or postponed orders.

The track length is 100 units and the Wagons report their positions every 10 seconds. There are three speed options available to the customer: low (corresponding to the default cruising speed of the Wagons, 0.3 units of length per second), medium (0.7 units of length per second) and high (1.1 units of length per second). On the considered track, the Warehouse Waiting Room (loading station) is at position 10, and the Test Station is at position 80 of the total length of 100 units. When loaded, the Wagons attempt to keep the requested speed as long as crashes are avoided, but on the return track (from Test Station back to the Loading Station), they revert to the default speed, considered most energy-efficient.

The exogenous parameters refer to the expected number and distribution of orders. Using the Overture (VDM-RT) HMI allowed for a good control of both parameters. The orders have been generated from the Uniform and Normal (Gaussian) distributions. The Gaussian distribution is justified by the Central Limit Theorem when the demand comes from many different independent or weakly dependent customers, whereas the uniform distribution could be the result of orders coming from independent customers at various time zones (or evenly populated meridians).

The number of orders over a 12 hour period was selected between 200 and 300 orders, about half of which would be cancelled if the order does not reach the end user within 900 seconds (15 min) from the request. For the uniform distribution, the orders are expected to come every 144 s (for 300 orders) to 216 s (for 200 orders) on average. The actual sets of orders and their distribution for the extreme cases (200 and 300 orders), as generated with the MATLAB® random variable functions, are shown in Fig. 7.10.

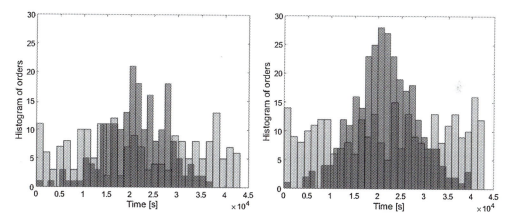

Figure 7.10 Histogram of orders for 200 (left) and 300 (right) orders from the Gaussian (blue; mid grey in print version) and Uniform (orange; light grey in print version) distributions.

Additionally, one order (an all–black stick with requested speed "low" and no cancellation) is inserted in the list at the start, to initialise the system. These sets of orders are used to test the load and overload of the production line. The other parameters of the orders are fixed:

- the colour choices are random from the discrete uniform distribution
- the speed options are also uniformly distributed (about a third of the orders contain each option)

The endogenous parameters considered in the current analysis are:

- The number of available colours
- The size of the Warehouse memory box
- The transport capability of Wagons

The number of available colours has been chosen between 4, 6 and 8, considering that fewer than 4 colours is not an interesting test, and that more than 8 colours is impossible without re-defining the concrete message encoding between subsystems. Having a maximum of 8 available colours (limited due to the structure of the messages between HMI and Part Tracker, as well as between Part Tracker and various other units such as the Warehouse and the Test Station), the lowering of the colour set may have a positive impact on the time an order spends in the Warehouse before being assembled.

The size of the Warehouse memory box could also be adjusted between 10 slots for each part type up to 60 slots for each part type. Unlike the colours, the memory box size is internal to the Warehouse and can be adjusted freely. However, large memory boxes may be impractical (may use a lot of space) or undesirable by the producer.

The transport capability of the wagons refers to the number of units that can be loaded on a Wagon. The capacity of each Wagon can be maximum 2 sticks at a time. Of course, wagons can still carry a single stick, if a second one is not readily available in the Waiting Room next to the Warehouse. The testing time of the sticks in the Test Station is fixed to 16 s per unit and the test result has a theoretical 2% probability of rejection, in which case the order returns to the Warehouse for re-assembly.

Cost functions

The improvements of the setups have been quantified by the following cost functions which would ideally all be at a minimum at the same time.
- CF0: Percentage of cancelled orders
- CF1: Average Lead Time
- CF2: Number of discarded parts

The percentage of cancelled orders is the most important cost function from a practical point of view in a real production line, since cancelled orders become unsold items. Only about half of the orders have cancellation policies (if the order is not received in 900 s from submitting the request). Lead time is considered the length of time between the order placement and the moment the system delivers the requested unit to the end user. The number of parts discarded for recycling is an important indicator of both the Warehouse lead time and the memory box overload. While these measures have been computed separately, keeping the number of discarded parts low improves both.

Analysis when varying exogenous parameters

For the first round of experiments, the size of the memory box (10/25/40 slots) and the external parameters (order distribution: uniform/Gaussian, and number of orders: 201/251/301) are varied to generate 9 experiments per distribution. The uniform distribution experiments do not produce cancelled orders. Even with the highest number of orders tried, the Uniform distribution rarely goes over 14 orders in 1500 s (25 minutes), whereas the Gaussian distributions concentrate at least 20 orders in the most loaded 25-minute timeslot, preceded and followed by other crowded timeslots. The number of orders over the whole 12 h experiment is therefore less indicative of the number of cancellations than the load in the peak timeslots. The system can handle constant flows of 10–15 orders per 25-minute timeslot, but when one timeslot goes over 20 orders, the system is overloaded, and the delays pile up and negatively affect following orders.

Analysis when varying the number of memory box slots

The adjustment of the number of memory box slots makes more sense for the experiments when the system is close to be overloaded by orders. The most important effect of increasing the memory box size is the reduction of discarded parts (due to the memory box being large enough to usually hold enough samples of all colours) shown in Fig. 7.11. However, even this effect is significantly diminishing for memory boxes with more than 50 slots. The lead time is decreasing due to the lower time spent in or before the Warehouse. But because the Warehouse time is less significant than the waiting time after this stage (due to Wagons being elsewhere on the track), the decrease in lead time is rather mild. The number of cancelled orders is not affected as much by this setting, as all memory box sizes over 10 reduce the cancelled order percentage by about 10%.

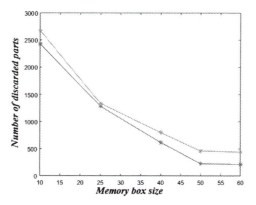

Figure 7.11 Evolution of the number of discarded parts as a function of the memory box size, for 301 orders from the Gaussian (blue; dark grey in print version) and Uniform (orange; light grey in print version) distributions.

Analysis when doubling the load capacity of the Wagons

The doubling of the Wagon capacity reduces the lead time primarily by reducing the waiting time after the Warehouse (to approximately a third) and therefore the average lead time is reduced significantly enough for eliminating almost all cancellations. Doubling the carrying capacity has no drawback and it allows for lower sizes of the memory box to achieve a low number of discarded parts, as will be tested in the next set of experiments.

Analysis when varying the number of colours available to the user

Reducing the number of colours should have a beneficial effect on the lead time by increasing the probability that a requested part is available in the memory boxes. The effect of reducing the number of colours will be analysed in conjunction with the dimension of the memory box, since both operations aim to improve the Warehouse time and reduce the number of discarded parts. The point of interest for this set of analyses is the balance between the two parameters, in the sense of identifying the amount by which the memory box can be downsized when reducing the number of available colours.

Table 7.4 shows the lead time when both parameters (number of colours available to the user and size of the memory box) are varied. For almost all experiments (except Gaussian distribution with a memory box of 10 slots), at most 2 orders were not fulfilled, making the percentage of cancellations reasonable (under 1%). Regarding the lead time, it is not significantly decreasing with a memory box of 25 slots or more, regardless of the number of colours. The number of discarded parts does not decrease as sharply starting with 40 slots in the memory box, while still being higher for larger numbers of available colours. Therefore, a good balance is achieved when using 6 colours and 40 slots. The average time the order spends at each stage for these experiments is shown in Fig. 7.12.

Table 7.4 Average lead time [s] and number of discarded parts for various combinations of the number of available colours and the size of the memory box.

Uniform Distribution		Size of the memory box			
Number of available colours		50 slots	40 slots	25 slots	10 slots
	8	288.56s	286.49s	310.75s	360.56s
		691 parts	625 parts	1450 parts	2696 parts
	6	274.35s	273.75s	284.04s	300.60s
		216 parts	325 parts	758 parts	1536 parts
	4	260.31s	266.11s	263.60s	289.67s
		102 parts	103 parts	149 parts	600 parts

Gaussian Distribution		Size of the memory box			
Number of available colours		50 slots	40 slots	25 slots	10 slots
	8	330.07s	379.07s	387.12s	450.46s
		416 parts	664 parts	889 parts	2065 parts
	6	315.38s	356.74s	313.03s	417.53s
		171 parts	180 parts	544 parts	1340 parts
	4	293.20s	332.17s	309.54s	319.22s
		20 parts	129 parts	230 parts	475 parts

7.6.2 Analysis of the homogeneous simulation in relation to the physical system

Compared to the real system, the VDM-RT models of the homogeneous co-simulation are not meant to be 100% accurate in the physical (mechanical / electrical) implementation sense. The purpose of the VDM-RT models is to capture abstractly the interfaces between the constituent subsystems. The incompleteness of the VDM-RT models is related to details of the inner workings of the components, not necessarily respecting all the constraints of reality. For instance, the inexhaustible stacks of randomly generated colours for the USB-OTG parts in the Warehouse or the fixed testing time of 16 s per unit are clearly simplifying assumptions. However, the homogeneous co-simulation is relatively predictive in the sense that an increase or decrease of the average lead time, computed statistically over a sufficiently large number of orders, would be reflected in

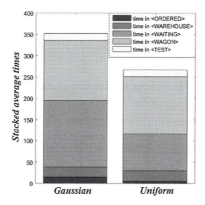

Figure 7.12 Stacked average times at each stage of production for the considered set of parameters.

an increase or decrease of the average lead time on the real system, regardless of the actual inaccuracies. Warehouse operations such as moving a part to/from a memory box or moving an assembled stick to the Waiting Room may take slightly more or less than the four seconds of the simulation, but what would decisively influence the average lead time is the number of such operations. The focus during the homogeneous co-simulation phase was to improve system-level parameters rather than achieving perfect predictions. Furthermore, considering that the VDM-RT models have been later replaced with models generated in specialised tools, the real system performance was matched against the latter heterogeneous co-simulation rather than the homogeneous one.

7.6.3 Heterogeneous phase experiments

The heterogeneous co-simulation covers the successful attempts to replace the Overture-generated FMUs with more detailed and complete FMUs generated by specialised tools using appropriate formalisms. Table 7.3 shows the correspondence between the subsystems and the adequately suited program for implementing a complete simulation for each.

Having already established a homogeneous co-simulation (with all FMUs generated from Overture), the improvements to various parts of the project can be achieved independently from each other, because any newly generated FMU would simply replace the corresponding Overture/VDM-RT FMU if the interfaces between subsystems remained unchanged. Thus, the order of integration for completely-functional FMUs is determined by the progress of individual teams rather than pre-defined dependencies. Teams were able at any time to rely on a working DE-first co-simulation and only replace their unit. Furthermore, the co-simulation could at any time be assembled from any combination of FMUs (generated by Overture or other tools).

While for most subsystems the FMU generated by the specialised program is simply an improvement (a more realistic or more detailed simulation) over the Overture model, the test case captures two exceptions: the HMI and the Part Tracker.

The HMI is different from the rest of the subsystems in the sense that the two FMUs are meant to complement each other, although they cannot be used both at the same time. The Overture/VDM-RT FMU reads the orders from a *.csv file, and has been used for performing benchmark like tests with completely controlled and repeatable sequences of orders. The 4DIAC+MQTT FMU allowed for real-time placement of orders and real user interaction with the simulation. Furthermore, because orders can be requested from a smartphone or tablet, the 4DIAC+MQTT FMU also implements a graphical interface.

The Part Tracker handles the flow of the process, it is a logical unit rather than a physical one. Therefore, the Part Tracker is the only unit whose FMU is generated in Overture throughout the heterogeneous co-simulation phase, also being the only truly complete VDM model.

For one of the other units, the flexibility of the co-simulation allowed for a successful implementation of a back-up plan. The FMU for the Robotic Arm was initially intended to be produced in Catia v6, but the project actually used an alternative 20-sim FMU model.

The co-simulation in the heterogeneous phase allows for the analysis of interactions between the units that have been simulated using specialised programs. All messages exchanged between VDM FMUs (or their more rigorous versions) are available for display in the co-simulation engine and INTO-CPS Application. For instance, the progress of assembly on the orders is clearly noticeable in Fig. 7.13 because acknowledgement messages are sent back to the HMI whenever the item reaches the next stage. Furthermore, the acknowledgement messages contain the ID of the order, which would make the process easy to follow even in case of items being assembled in parallel, as if in a pipeline. Another example is shown in Fig. 7.14 with the monitoring of the positions and frontal distances reported by the Wagons.

The subsystems modelled and tested by the heterogeneous co-simulation have been deployed in a demo stand (Fig. 7.2) for fine tuning under real-life conditions. The HMI is a virtual unit, which was the first to be implemented on smartphones and tablets as an app, communicating with the Part Tracker through MQTT even during the heterogeneous phase of the project.

The Part Tracker is a logical unit for which the FMU was generated by the Overture tool in both the homogeneous and the heterogeneous phases. For deployment, the C code was generated directly from Overture and deployed on a Raspberry Pi 3, also employing MQTT as communication protocol [28].

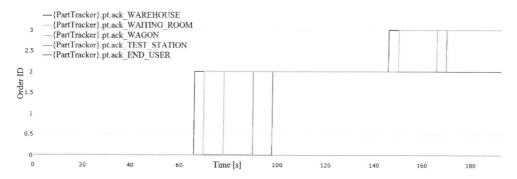

Figure 7.13 Acknowledgement messages for stages of assembly.

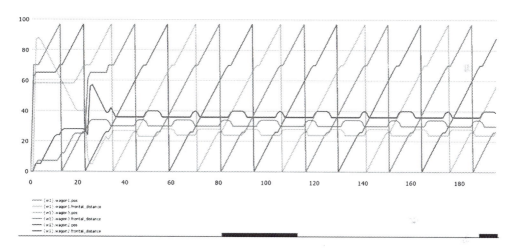

Figure 7.14 Wagon positions and frontal distances.

The Warehouse, mainly consisting of stacks and memory locations, colour sensors and pneumatic pistons, uses a Raspberry Pi 3 with UniPi Expansion Board for deploying the 20-sim generated C code of the heterogeneous co-simulation.

The Robotic Arm is a Stäubli robot with no internal logic, just following the lead of the Warehouse.

The Wagons, being the moving parts of the stand, contain DC motors (with PWM drivers), position sensors for detecting the loading and testing station, anti-collision ultrasonic sensors to avoid collision with other Wagons and embedded Raspberry Pi boards for the internal logic generated from the 4DIAC FMU.

Finally, the testing station uses a camera for image processing and also includes actuators to push the USB sticks from the Wagons into the pile of rejected items or towards the end user.

7.7. Conclusion

The chosen case study of producing USB-OTG sticks offers a balance between being sufficiently simple to be easily followed as a production line example, including a tangible output, and at the same time being sufficiently general to allow the study of the co-simulation complexity, as well as the exploration of the capabilities and limitations of both the methodology and the tools.

7.7.1 Two-phase development

Each subsystem was first modelled abstractly in VDM-RT, using the Overture tool. The goal of this homogeneous co-simulation was twofold: to identify the right interaction protocols (signals) among the various components (stations) of the prototype and to identify conceptual (system-level) parameters before the design space exploration at the heterogeneous co-simulation step. The homogeneous phase experiments were performed under the assumptions that the orders are received in a Gaussian or Uniform distribution, with balanced distributions of colours for each side of the stick and balanced requested Wagons speeds. Two types of analysis have been performed:

- Analysis of the behaviour of the system when varying exogenous parameters, to identify the maximum load that the system is capable of handling.
- Analysis and optimisation of the system when varying endogenous parameters, to find a balanced set of parameters.

The analysis on internal parameters allowed the balanced decision for the system-level parameters. These parameters have been taken into account and reduced by a considerable amount the design space exploration at the heterogeneous stage, in effect becoming system constraints for the heterogeneous simulation and the physical model.

Regarding the heterogeneous phase, the flexibility of the co-simulation engine allowed for the gradual integration of closer-to-reality and more detailed FMUs, generated in dedicated tools.

The most straightforward benefit in using the two-phase development is the possibility to simulate, test and validate (from a holistic perspective and with an increased level of accuracy) an entire production system that needs cross-functional expertise; in the past, these issues were tackled by experiential learning from multiple sequential implementations of the automation component; in other words, the co-simulation facilitates the adoption of agile software development principles for factory automation. For example, the initial development of a homogeneous co-simulation in VDM-RT for the iPP4CPPS prototype was particularly useful in driving cooperation and clarifying the assumptions of the teams involved in modelling specific components. Once the VDM-RT co-simulation was running, the independent development of units were integrated, validated and deployed when ready, since the methodology allowed for their development in any order.

Another benefit is the ability to handle unpredictable events, to a certain extent. The use of co-simulations when designing an automated production system avoids the build-up inertia of subsequent design constraints, facilitating the low and late commitment for these decisions, like the specific micro-controllers or PLCs, as well as allowing the change of modelling tools for any subsystem at any time during the project.

In addition, the two-stage approach is applicable to other CPS modelling endeavours, especially when system-level parameters can be identified and adjusted with abstract models for most, or even all, of the subsystems. As an example, selecting the suitable wireless connectivity in Industry 4.0 is determined by parameters such as latency, reliability and throughput.

Having an early simulation that can give communication patterns and strains, including for instance the number, size and composition of messages, can aid in the selection of the communication and processing units, as well as the identification of critical data that need to be sent/received synchronously. Conversely, if the communication network is fixed or can only be extended at great cost, the early co-simulation could give a rough estimate or upper bound on the capacity of the system.

The methodology can be naturally extended to include interactions between generated hardware and generated software solutions, which in the case of the production line may refer to writing data onto the USB-OTG sticks and verifying it as part of the production process.

7.7.2 Methodological insights

From the methodological standpoint the experiment provides several insights from employing the model-based approaches to build complex CPS-based manufacturing systems. Complex co-simulations for production systems require one to rely on more mature methodologies with relevant track-records in the domain, such as agent-oriented and component-based. Even if both methodologies (agent-oriented and component-based) promote the classical features of agile software development (iterative, incremental, lightweight and collaborative) they use different abstractions that make them suitable to describe either the conceptual or the implementation model.

While the agent-oriented approach provides the most adequate abstractions to design the conceptual model of the co-simulation, some of its specifics make it difficult to implement in the available co-simulation technology which is more related to component-based approaches.

In the experiments, the agent-oriented approach has been used to identify the subsystems and to structure the messages among the specific co-simulation units, whilst its implementation has followed the component-based approach. This was needed for at least two reasons:

- The components' meta-model does not have abstractions for the goals and components can only use task delegation. In manufacturing systems there are multiple conflicting objectives at both system level (e.g. throughput, lead time, etc.) and individual level (e.g. energy consumption, degree of utilisation, slack time, etc.). The right balance among all these conflicting objectives is either unknown or underspecified and is usually discovered during the design space exploration phase. Any predefined way of achieving a goal (task delegation) may inhibit optimisations. While not tested in the experiment, the possibility to allow internal optimisation is available in subsystems, for instance in the Warehouse or the Test Station. The Warehouse receives information about the requested colours of a USB-OTG stick, but it is not forced to look for the necessary parts in any specific order. The VDM-RT model uses the order Left–Middle–Right because of the bottleneck of having a single Robotic Arm capable of moving parts to and from the memory boxes. The usage of two Robotic Arms, for example, would introduce an optimisation problem where the Warehouse would have to decide which side to start from. Consequently, goal delegation may be implemented in the current co-simulation technology.

- For agent-oriented approaches, the environment is a first-class abstraction that is a structural part of an agent's meta-model, while it is not part of a component's meta-model. In manufacturing the infrastructure is part of any referenced architecture and therefore special attention has been given to model and simulate the Part Tracker. The interaction with the environment is different, agents can measure the environment, while components can react to an event only by defining a relation with it. A feasible extension considering the current co-simulation technology would therefore be, in the case of multiple Warehouse units, to allow the Wagons to take the decisions themselves whether to wait longer in the loading station of a particular Warehouse (even if there are no sticks currently in the corresponding waiting room) or to leave for another loading station (for instance in case another Wagon is close enough). In this scenario, the Wagon would acquire by itself the information about the existence (or not) of ready USB-OTG sticks in the waiting rooms (e.g. by enquiring the Part Tracker) and would have to be aware of the loading process, would have to decide on which slot to place the sticks and would know when the loading is complete.

The homogeneous-heterogeneous method discussed in this chapter is, in the context of a multi-paradigm co-simulation tool, a powerful and flexible approach for the development of CPS production lines, in industries such as manufacturing and product development.

Acknowledgements

The current work is partially supported by the European Commission as a small experiment selected for funding by the CPSE Labs innovation action (grant number 644400). The work presented here is also partially supported by the INTO-CPS project funded by the European Commission's Horizon 2020 programme (grant agreement number 664047). We would like to thank all participants of both projects for their contributions making the results reported here possible.

References

[1] Cláudio Gomes, et al., Co-simulation: a survey, ACM Computing Surveys (ISSN 0360-0300) 51 (3) (May 2018) 49 (pp. 1–33).

[2] Mihai Neghina, et al., Multi-paradigm discrete–event modelling and co-simulation of cyber-physical systems, Studies in Informatics and Control 27 (1) (2018) 33–42.

[3] Mihai Neghina, Constantin-Bala Zamrescu, Ken Pierce, Early-stage analysis of cyber-physical production systems through collaborative modelling, Software and Systems Modeling 19 (3) (2020) 581–600.

[4] IPP4CPPS, Integrated product-production co-simulation for cyber-physical production system, http://centers.ulbsibiu.ro/incon/index.php/ipp4cpps/. (Accessed June 2019).

[5] M. Mühlhäuser, Smart products: an introduction, in: M. Miihlhauser, A. Ferscha, E. Aitenbich (Eds.), Constructing Ambient Intelligence, Springer, Berlin, Heidelberg, 2008.

[6] Imran Quadri, et al., Modeling methodologies for cyber-physical systems: research field study on inherent and future challenges, Ada User Journal 36 (4) (2015) 246–253, http://www.ada-europe.org/archive/auj/auj-36-4.pdf.

[7] Marcel Verhoef, Peter Gorm Larsen, Jozef Hooman, Modeling and validating distributed embedded real-time systems with VDM++, in: Jayadev Misra, Tobias Nipkow, Emil Sekerinski (Eds.), FM 2006: Formal Methods, in: Lecture Notes in Computer Science, vol. 4085, Springer–Verlag, 2006, pp. 147–162, https://doi.org/10.1007/11813040_11.

[8] C.B. Zamfirescu, et al., Preliminary insides for an anthropocentric cyber-physical reference architecture of the smart factory, Studies in Informatics and Control 22 (3) (2013) 269–278.

[9] Design Principles for Industrie 4.0 Scenarios, IEEE, ISBN 9780-7695-5670-3, 2016, https://doi.org/10.1109/HICSS.2016.488.

[10] Peter Gorm Larsen, et al., The overture initiative integrating tools for VDM, SIGSOFT Software Engineering Notes (ISSN 0163-5948) 35 (1) (2010) 1–6, https://doi.org/10.1145/1668862.1668864, http://doi.acm.org/10.1145/1668862.1668864.

[11] John Fitzgerald, et al., Cyber-physical systems design: formal foundations, methods and integrated tool chains, in: FormaliSE: FME Workshop on Formal Methods in Software Engineering, ICSE 2015, Florence, Italy, 2015.

[12] Peter Gorm Larsen, et al., Integrated tool chain for model-based design of cyber-physical systems: the INTO-CPS project, in: 2016 2nd International Workshop on Modelling, Analysis, and Control of Complex CPS (CPS Data), IEEE, Vienna, Austria, 2016, http://ieeexplore.ieee.org/document/7496424/.

[13] Peter Gorm Larsen, et al., Collaborative modelling and simulation for cyberphysical systems, in: Trustworthy Cyber-Physical Systems Engineering, Chapman and Hall/CRC, ISBN 9781498742450, 2016.

[14] Peter Gorm Larsen, et al., Towards semantically integrated models and tools for cyber-physical systems design, in: Tiziana Margaria, Bernhard Steffen (Eds.), Leveraging Applications of Formal Methods, Verification and Validation, Proc. 7th Intl. Symp., in: Lecture Notes in Computer Science, vol. 9953, Springer International Publishing, ISBN 978-3-319-47169-3, 2016, pp. 171–186.

[15] John Fitzgerald, et al., Collaborative model-based systems engineering for cyber-physical systems - a case study in building automation, in: Proc. INCOSE Intl. Symp. on Systems Engineering, Edinburgh, Scotland, 2016.

[16] T. Blochwitz, et al., The functional mockup Interface 2.0: the standard for tool independent exchange of simulation models, in: Proceedings of the 9th International Modelica Conference, Munich, Germany, 2012.

[17] Casper Thule, et al., Maestro: the INTO-CPS co-simulation framework, Simulation Modelling Practice and Theory (ISSN 1569-190X) 92 (2019) 45–61, https://doi.org/10.1016/j.simpat.2018.12.005, http://www.sciencedirect.com/science/article/pii/S1569190X1830193X.

[18] Softeam, Modelio, https://www.modelio.org/. (Accessed June 2019).
[19] Christian Kleijn, Modeling and simulation of fluid power systems using 20-sim, International Journal of Fluid Power 7 (3) (2006) 5760.
[20] Linköping University, OpenModelica official website, http://www.openmodelica.org/, 2015.
[21] T. Strasser, et al., Framework for distributed industrial automation and control (4DIAC), in: 2008 6th IEEE International Conference on Industrial Informatics, 2008, pp. 283–288, https://doi.org/10.1109/INDIN.2008.4618110.
[22] 3DS, CATIA, https://www.3ds.com/products-services/catia/. (Accessed June 2019).
[23] John Fitzgerald, Peter Gorm Larsen, Marcel Verhoef (Eds.), Collaborative Design for Embedded Systems – Co-modelling and Co-simulation, Springer, 2013.
[24] D. Bjørner, Programming in the meta-language: a tutorial, in: The Vienna Development Method: The Meta-Language, 1978, pp. 24–217.
[25] John Fitzgerald, Peter Gorm Larsen, Modelling Systems - Practical Tools and Techniques in Software Development, Cambridge University Press, The Edinburgh Building, Cambridge CB2 2RU, UK, ISBN 0-521-623480, 1998, https://doi.org/10.1145/1668862.1668879.
[26] P. Leitão, S. Karnouskos (Eds.), Industrial Agents. Emerging Applications of Software Agents in Industry, Elsevier, 2015.
[27] RAMI, Referenzarchitekturmodell Industrie 4.0, DIN SPEC 91345:201604, https://www.plattform-i40.de/I40/Redaktion/DE/Downloads/Publikation/din-spec-rami40.html. (Accessed June 2019).
[28] Victor Bandur, et al., Code-generating VDM for embedded devices, in: John Fitzgerald, Peter W.V. Tran-Jørgensen, Tomohiro Oda (Eds.), Proceedings of the 15th Overture Workshop, Newcastle University, 2017, pp. 1–15, Computing Science. Technical Report Series. CS-TR-1513.

CHAPTER 8

Agent-based cyber-physical system development with SEA_ML++

Moharram Challenger[a], **Baris Tekin Tezel**[b,c], **Vasco Amaral**[d], **Miguel Goulão**[d] and **Geylani Kardas**[b]

[a]University of Antwerp and Flanders Make, Antwerp, Belgium
[b]International Computer Institute, Ege University, Izmir, Turkey
[c]Dokuz Eylul University, Izmir, Turkey
[d]University of Lisbon, Lisbon, Portugal

8.1. Introduction

According to Russell and Norvig [1], an agent is anything that can be considered to be able to perceive its environment through sensors and act on this environment through actuators. Moreover, the agents are located in a certain environment and are capable of flexible autonomous actions within this environment in order to meet its design objectives [2]. These autonomous, reactive, and proactive agents also have social ability and can interact with other agents and humans in order to solve their own problems. They may also behave in a cooperative manner and collaborate with other agents for solving common problems.

In order to perform their tasks and interact with each other, intelligent agents constitute systems called Multi-agent Systems (MASs). A MAS is a loosely coupled network of problem-solving entities (agents) that work together to find answers to problems that are beyond the individual capabilities or knowledge of each entity (agent).

Agents and MASs with their capabilities such as mobility, intelligence, distributedness, autonomy and dynamicity, can be utilised in very different applications, ranging from software intensive applications such as e-bartering [3,4] and the stock exchange system [5], to the system level applications such as smart waste collection [6]. MASs can also be used along with other paradigms such as Model-based System Engineering (MBSE) to target the challenges of Cyber-physical Systems (CPS) including resource limitation, uncertainty, and distributedness. Hence, the fundamental components of CPS for various business domains can be designed and built as autonomous agents interacting with each other. MBSE is here applied to leverage the abstraction level to minimise the system complexity and facilitate the agent design. Finally, it provides a convenient way to implement and execute these systems for various MAS execution platforms [7–9].

This chapter discusses how agents and MAS can be used in both the modelling and the development of CPS. To this end, we demonstrate how using a Domain-specific

Multi-Paradigm Modelling Approaches for Cyber-Physical Systems
https://doi.org/10.1016/B978-0-12-819105-7.00013-1

Modelling Language (DSML), called SEA_ML++ enables the model-driven engineering of agents, their planning mechanisms and agent collaborations which leads to the construction of the desired CPS. SEA_ML++ can represent different aspects of MASs such as environment, interaction, agent internal, organisation, plan, and role [10]. In this way, it can specify various aspects of a complex and dynamic system such as CPS. To demonstrate the proposed development methodology, a case study called multi-agent garbage collection is designed and implemented based on SEA_ML++.

This chapter is organised as follows: Sections 8.2 and 8.3 discuss the background and related work, respectively. In Section 8.4, SEA_ML++ is introduced. The methodology for modelling and development of agent-based CPS is discussed in Section 8.5. A multi-agent garbage collection system is designed and developed using SEA_ML++ in Section 8.6 to demonstrate the proposed methodology. Finally, the chapter is concluded in Section 8.7.

8.2. Background

Cyber-Physical Systems (CPS) consist of tightly integrated and coordinated computational and physical elements [11]. They represent an evolution of embedded systems to a higher level of complexity by focusing on the interaction with highly uncertain environments (such as human interaction or wear & tear of devices). In these systems, embedded computers and networks monitor (through sensors) and control (through actuators) the physical processes, usually with feedback loops where physical processes and computations affect each other. The computational part of these systems plays a key role and needs to be developed in a way that can handle (mostly in real-time) the uncertain situations with the limited resources (including computational resource, memory resource, communication resource, and so on) [12]. However, considering both the heterogeneity of the components and the variety of system behaviour interacting with physical environment, the design and the development of these systems are complex, time-consuming and costly tasks.

Generally, one of the approaches to address the complexity of engineering systems is to exclude the extra details and have an abstract model/representation of the system where we can do some tasks (e.g. analyse, comprehend and develop) which are difficult or sometimes impossible to do on the original system [13]. Modelling a system represents the properties of interest in that system which can be used for various purposes. There can be different models (with specific paradigms or formalisms) for a complex system such as a CPS in which each of the models represents one aspect of the system. This approach is called multi-paradigm modelling [14]. The modelling approach can be used for different purposes, such as model-driven engineering which is a software and systems development paradigm that emphasises the application of modelling principles and best practices throughout the System Development Life Cycle (SDLC).

Within an MDE approach, a DSML uses the notations and constructs tailored towards a particular application domain (e.g. MAS or Concurrent Programs [15]). The end-users of DSMLs have knowledge from the observed problem domain, but they usually have little programming experience. DSMLs raise the abstraction level, expressiveness, and ease of use. The main artifacts of DSML are models instead of software codes and they are usually specified in a visual manner. A DSML's graphical syntax offers benefits, like easier design, when modelling within certain domains (e.g. IoT domain [16–18]).

The development of DSML is usually driven by language model definition. That is, concepts and abstractions from the domain which need to be defined in order to reflect the target domain (language model). Then relations between the language concepts need to be defined. Both of them constitute an abstract syntax of a modelling language. Usually, a language model is defined with a metamodel. The additional parts of a language model are constraints that define those semantics which cannot be defined using only the metamodel. Domain abstractions and relations need to be presented within a concrete syntax and serve as a modelling block within the end-user's modelling environment. This modelling environment can be generated automatically if dedicated software is used, otherwise, modelling editor must be provided manually. Then the model transformations need to be defined in order to call the domain framework, which is a platform that provides the functions for implementing the semantics of DSMLs within a specific environment. Usually, the semantics is given by translational semantics.

8.3. Related work

This section discusses the related work considering both the MAS DSMLs and the Agent-based CPS development studies.

Agent-DSL [19] is used for modelling agent features, like knowledge, interaction, and autonomy by presenting a metamodel. The agent modelling languages introduced e.g. in [20,21] consider the syntax definitions rather than operational language semantics. Studies like [22,23] also discuss MDE of agent systems by introducing a series of transformations on MAS metamodels in different abstractions. Although those transformations may guide to construct some sort of semantics, related studies describe MAS development methodologies instead of specifying a complete DSML. In addition, there exist MAS metamodel proposals (e.g. [24,25]) from which abstract syntaxes of MAS DSMLs can originate. In [26], a DSML is provided for MASs based on EMF.[1] The language supports modelling of agents according to one of the specific MAS methodologies called Prometheus. Likewise, SEA_L [27] and JADEL [28] are two agent DSLs both

---

[1] Eclipse Modeling Framework, http://www.eclipse.org/modeling/emf/.

providing textual syntaxes based on Xtext specifications. Sredejovic et al. [29] introduce another agent DSL, called ALAS, to allow software developers to create intelligent agents having reasoning systems based on non-axiomatic logic. The work conducted in [20] aims at creating a UML-based agent modelling language, called MAS-ML, which is able to graphically model various types of agent internal architectures. However, the current version of the language does not support any code generation, which prevents the execution of modelled agent systems. DSML4BDI [9] is another DSML proposed for creating agents conforming to Belief-Desire-Intention (BDI) architecture. In addition to modelling the internal structure of agents, their beliefs, goals, events and knowledgebase, DSML4BDI specifically allows modelling the difficult logical expressions, which might be used in any agent plan or rule. In addition to providing an abstract syntax based on a metamodel, SEA_ML++ [10,30] offers a full-fledged modelling language including all syntax and semantics constructs required for the MDE of agents according to well-known BDI and re-active agents principles. SEA_ML++ supports the execution of modelled agents over a series of model-to-model and model-to-code transformations enabling the construction of interactions between agents.

In Road2CPS EU support Action,[2] a roadmap and recommendations for future deployment of CPS are proposed [31]. Similarly, in CPSoS EU Support Action,[3] the challenges posed by engineering and operation of CPSoS are defined [32,33] and a research and innovation agenda on CPSoS are presented. In [34], the authors report recent software engineering studies for CPSs. Also, in [35] and [36], the authors address multi-paradigm modelling aspect of cyber-physical systems. Finally, in [37], the authors present a DSL for designing CPSs. However, none of these studies address the construction of these systems using agents and multi-agent systems, e.g. to provide autonomy, reactivity and/or proactivity.

The study in [38] addresses the modelling methodology and tool for autonomous objects in CPS. The authors propose a framework called CPS-Agent to model objects with consideration of temporal–spatial traits and interaction with the physical environment. They present a role-based strategy formulation to make work patterns of CPS-Agents more clear. In terms of network communication among CPS-Agents, a set of communicative primitives is tailored based on the FIPA-ACL specification.

In [39], the authors discuss the development of an agent-based CPS for Smart Parking Systems. They believe that the inclusion of MAS, combined under the scope of CPS, ensures flexibility, modularity, adaptability and the decentralisation of intelligence through autonomous, cooperative and proactive entities. They also mention that the smart parking systems can be adapted to other types of vehicles to be parked and scalable in terms of the number of parking spots and drivers/vehicles. The authors focus

---

[2] http://www.road2cps.eu/.
[3] https://www.cpsos.eu/.

on how the software agents are interconnected with the physical asset controllers using proper technologies.

The authors of [40] address the challenges of MAS for CPS. They state that CPS application domains include three major characteristics (intelligence, autonomy and real-time behaviour) and MAS can be used to implement such systems. They believe that MAS address the first two characteristics, but miss to comply with strict timing constraints. The main reasons for this lack of real-time satisfiability in MAS originate from current theories, standards, and technological implementations. In particular, internal agent schedulers, communication middlewares, and negotiation protocols have been identified as co-factors inhibiting the real-time compliance.

Recently in [41,42], a group of researchers (in the scope of a European research project[4]) have studied the application of agents to deal with the problems in Cyber-physical Production Systems. They focused on quality control in manufacturing by proposing an agent architecture (which presents a distributed intelligence) for distributed analysis of the CPS and tackling the defects in multi-stage manufacturing.

Although all of the above-mentioned studies present noteworthy applications of agent paradigm in CPS, they do not address the MDE of these systems in the way of both increasing the abstraction level needed during CPS design and facilitating CPS construction with agent features. The enrichment of CPS implementations with the agent capabilities improves the execution of such systems which may also lead to the widespread use of CPS in various application domains e.g. ranging from smart manufacturing to self-adaptive systems. However, design and implementation of CPS with agent capabilities naturally become more difficult in comparison with the conventional way of CPS development since the developers may face new challenges originating especially from the autonomous and the proactive behaviour of the agents built in the new CPS in addition to the already existing CPS development challenges such as the complexity and the interoperability of CPS components. MDE of agent-based CPS may contribute to minimise these challenges and hence provide a more convenient way of development. Within this context, this chapter investigates the use of a DSML enabling the MDE of CPS which is missing in the existing agent-based CPS development approaches.

8.4. SEA_ML++

This section elaborates the SEA_ML++ language and its components, including its abstract syntax, graphical concrete syntax, and transformations.

SEA_ML++ is an extended version of SEA_ML [43] and SEA_L [27,44] languages by the systematic evaluation [7,30] of agent modelling components and applying the physics of notation principles [45] to improve the graphical syntax [10] used during

---

[4] http://go0dman-project.eu/.

MAS modelling. The initial idea for creating such a DSML is first introduced in [46] and its metamodel and concrete syntax in [24].

8.4.1 Abstract syntax

The abstract syntax of the SEA_ML++ language is constituted by a metamodel which is divided into different viewpoints each describing a different aspect of MAS. The important viewpoints are MAS/Organisation, Agent Internal, Plan, Role, Interaction, Environment, and Ontology. They were previously defined by the partial metamodels in the abstract syntax of the SEA_ML language [43]. However, all these partial metamodels were improved and combined into the metamodel of SEA_ML++.

SEA_ML++ covers the main agent entities and their relationship which are mostly agreed by the agent research community. In addition, more specific aspects of the domain, such as Plan, Role, and Environment, are also supported in the syntax of SEA_ML++ in detail. The general overview and relations between the viewpoints of SEA_ML++ are given in Fig. 8.1.

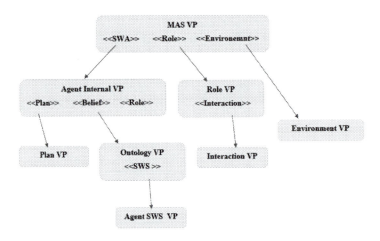

Figure 8.1 The general overview of the viewpoints of the SEA_ML++.

All viewpoints of SEA_ML++ are briefly discussed in the following:

MAS Viewpoint   The MAS viewpoint of SEA_ML++ is related to the creation of a MAS as an overall aspect of the metamodel. It contains the main blocks that form the complex system as an organisation.

Agent Internal Viewpoint   This viewpoint focuses on the internal structure of each agent in a MAS organisation. The abstract syntax of SEA_ML++ supports both reactive and BDI agents via this viewpoint. While meta-entities such as *Belief*, *Plan* and *Goal* support the BDI agents, *Behaviour* meta-entity and its nested structure support the reactive agents.

Plan Viewpoint   The Plan viewpoint defines the internal structure of an agent's plans. When an agent implements a *Plan*, it executes its *Tasks* that consist of *Actions* which are atomic elements. *Send* and *Receive* entities are extended from Action. These types of actions are linked to a Message entity.

Role Viewpoint   Agents can play some roles, and use ontologies, and infer about the environment based on the known facts within the system. The content of roles is defined in the Role viewpoint.

Interaction Viewpoint   This viewpoint focuses on communications and interactions between agents in a MAS, and identifies entities and relationships, such as *Interaction*, *Message*, *MessageSequence*.

Environment Viewpoint   The environmental viewpoint focuses on the relationships between agents and to what they access. The Environment in which the agents are located includes all non-agent entities too such as Resources (for example, a database, a network device), Facts and Services.

Ontology Viewpoint   A MAS Organisation can use ontologies for their reasoning. An ontology represents a source of information gathering and reasoning for all MAS members. All ontology sets and ontological concepts are brought together via the ontology viewpoint.

8.4.2 Graphical concrete syntax

A screenshot from the Sirius-based IDE of SEA_ML++ is given in Fig. 8.2. Developers can visually create models of agent systems conforming to SEA_ML++ specifications by simply drag-and-dropping the required items from the palette residing at the right side of the modelling environment.

8.4.3 Transformations

It is not enough to present a DSML simply by defining concepts and representations [9,15]. The exact definition entails the semantics of the language. The semantics of SEA_ML++ is given as a transitional semantics that is matching the concepts of the language in terms of the other concepts already established to realise MAS. Model-to-model transformations in SEA_ML++ are provided for MAS, Agent and Interaction models, to various MAS programming languages and their implementation platforms such as JADEX,[5] JACK,[6] JADE,[7] and Jason.[8] These agent programming languages were chosen as the target agent platform since they are among the well-known and

---

[5] https://sourceforge.net/projects/jadex/.

[6] https://aosgrp.com/products/jack/.

[7] https://jade.tilab.com/.

[8] http://jason.sourceforge.net/wp/.

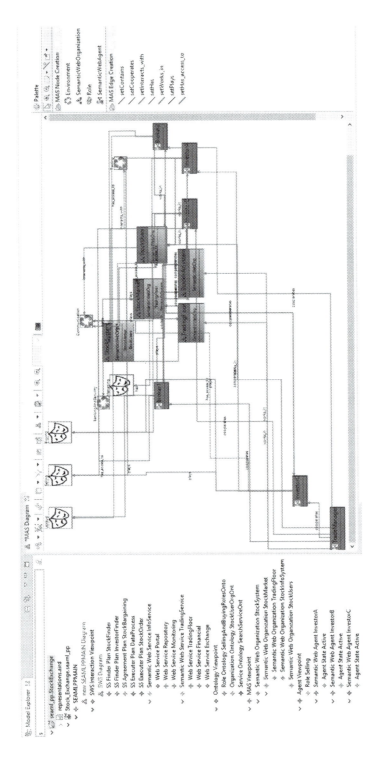

Figure 8.2 SEA_ML++'s IDE.

frequently used agent platforms in MAS research and development [47]. Also, JADEX, JACK and Jason support to develop BDI agents. In this study, ATL[9] translation language was used to provide model-to-model transformations between SEA_ML++ and above target platforms.

After the generation of platform-specific models through model transformations, a series of model-to-text transformations should be applied to generate executable software code for the modelled MAS. For this purpose, SEA_ML++ includes model-to-text transformation rules written in Acceleo to auto-generate MAS code from SEA_ML++ models.

8.5. Agent-based CPS modelling and development using SEA_ML++

In this section, a model-based methodology is proposed for the design and implementation of agent-based CPSs. To this end, the proposed methodology covers the use of the MAS DSML, SEA_ML++. In this way, a complex system such as CPS is modelled using MAS components at a higher level of abstraction. As a result, the system can be analysed, and the required elements can be designed using the terms and notations of agent and MASs. These domain-specific elements and their relations to each other create the domain-specific instance models which pave the way to implement the system. As these models are persisted in a structural and formal way, they can be transformed to other proper paradigms, such as mathematical logics. In this way, they can be analysed and validated based on formal methods, e.g. Satisfiability (SAT) solvers can be used to find counter-examples violating the agent model constraints (see [48] for a more extensive discussion). Furthermore, these models can be used to automatically generate the architectural code for agents and artifacts of the CPS which can end up with less syntactical errors and speed up the development procedure. Faster development also brings cost reduction in the projects. Moreover, less syntactic and semantic errors mean less iterations in the MAS development phase and short testing phase which also reduce the development cost and the effort. So, the MAS-to-be-implemented can be checked, and the errors can be partially found in the early phases of the development, namely analysis and design phases, instead of finding them in the implementation and testing phases.

In this chapter, SEA_ML++ is used as a DSML for the modelling and development of agent-based CPSs. SEA_ML++ enables the developers to model the agent systems in a platform-independent level and then automatically achieve code and related documents required for the execution of the modelled MAS on target MAS implementation platforms. In order to support CPS experts during MAS programming, SEA_ML++ covers all aspects of an agent system from the internal view of a single agent to the complex MAS organisation. In addition to these capabilities, SEA_ML++ also supports

---

[9] https://www.eclipse.org/atl/.

the model-driven design and implementation of autonomous agents who can work on CPS elements.

Based on SEA_ML++, the analysis and the modelling/design of CPS can be realised using the application domain's terms and notations. This helps the end users to work at a higher level of abstraction (independent of target platform) and close to the expert domain. Also, generative features of SEA_ML++ pave the way to produce the configured templates from the designed models for the software system in the underlying languages and technologies. Currently, SEA_ML++ can generate architectural code for several agent programming languages using model-to-model transformations of the designed platform-independent instance models to the instance models of the target MAS languages. Then these platform specific models are transformed to the platform specific codes by model-to-code transformations. This generation capability of SEA_ML++ can increase the development performance of the software system considerably. Finally, by constraints checking provided in SEA_ML++, the instance models are controlled considering domain-specific syntactic and semantic rules. These rules are applied in the abstract and the concrete syntaxes of the language. This feature helps to reduce the number of errors during the analysis and design of the software system and avoid postponing them to the development and the testing phases.

Although the new development methodology, introduced here, considers the adoption of SEA_ML++, it differentiates from the previous development approaches [4,43] brought for SEA_ML as being a complete development methodology for agent-based CPS, covering the analysis, design and implementation. Analysis and design phases include two types of iterations. Also another iteration loop is considered for the implementation and maintenance of the CPS. In addition to the modification of models, the methodology also supports the changes in auto-generated codes if required. The proposed SEA_ML++-based CPS development methodology includes several steps following each other (see Fig. 8.3): Agent-based CPS Analysis, Agent-based CPS Modelling, Automatic Code Generation, and finally code completion for exact agent-based CPS implementation.

Based on the proposed methodology, the development of an agent-based CPS starts with the analysis of the system by considering the MAS viewpoint of SEA_ML++ (see Fig. 8.3). This viewpoint includes MAS elements such as organisations, environments, agents and their roles. This viewpoint provides the eagle-view of the system and shapes the high-level structure of the system. The result is a partial platform-independent instance model of the system covering the analysis phase of the development and providing a preliminary sketch of the system.

In the system modelling step the CPS developer can use the fully functional graphical editors of SEA_ML++ to elaborate the design of the agents and MAS for the CPS under development, which can include seven viewpoints of the SEA_ML++'s syntax, in addition to the MAS viewpoint used in the analysis phase. These viewpoints cover all

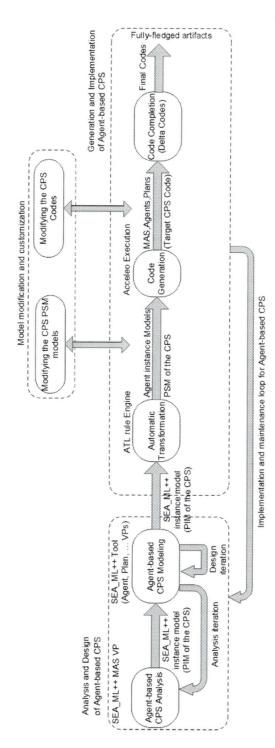

Figure 8.3 Agent-based CPS modelling and development using SEA_ML++.

aspects of a MAS. Each viewpoint has its own palette which provides various controls leading the designers to provide more accurate models. By designing each of these models for viewpoints, additional details are added to the initial system model provided in the analysis phase. These modifications immediately are updated in the diagrams of all other viewpoints. As the other viewpoints may have some constraint checks to control some properties related to the newly added element, the developer will be directed to complete those other viewpoints to cover the errors and warnings (coming from the constraint checks). This can lead to several iterations in the design phase. The result of this phase is the development of a complete and accurate platform-independent model for the designed MAS.

The next step in the agent-based CPS development methodology using SEA_ML++ is the automatic model transformations. The models created in the previous step need to be transformed from platform-independent level into the platform-specific level, e.g. to the Jason models as in the case study of this chapter. These transformations are called M2M transformations.

According to OMG's well-known Model-driven Architecture (MDA),[10] SEA_ML++ metamodel can be considered as a Platform Independent Metamodel (PIMM) and Jason metamodel can be considered as a Platform-specific Metamodel (PSMM). The model transformations between this PIMM and PSMM lead to the implementation of the agent-based CPSs on Jason agent execution platform. These transformations are implemented using ATL Language to produce the intermediate models which enable the generation of the architectural code for the agents and their artifacts. A CPS developer does not need to know the details of both these transformations written in ATL and the underlying model transformation mechanism. Following the creation of models in the previous modelling steps, the only action requested from a developer is to initiate the execution of these transformations via the interface provided by SEA_ML++'s Graphical User Interface (GUI).

Upon completion of model transformations, the developers have two options at this stage: 1) They may directly continue the development process with code generation for the achieved platform-independent MAS models or 2) if they need, they can visually modify the achieved target models (e.g. Jason models) to elaborate or customise them, which can lead to achieve more complete software in the next step, code generation. In either case, the outputs of this step are several system models each specific to a MAS execution platform (e.g. Jason platform).

The next step in the proposed methodology is the code generation for the MAS implementing the CPS. To this end, the developers' platform-specific models (conforming to PSMMs) are transformed into the code in the target languages. The M2T transformation rules are automatically executed on the target models and the codes are

---

[10] https://www.omg.org/mda/.

obtained for the implementation of the MAS. In SEA_ML++, it is possible to generate code for BDI agent languages such as Jason from SEA_ML++ models. Based on the initial models of the developer, the generated files and codes are also interlinked during the transformations where they are required. To support the interpretation of SEA_ML++ models, the M2T transformation rules are written in Acceleo. Acceleo is a language to convert models into text files and uses metamodel definitions (Ecore files) and instance files (in XMI format) as its inputs. More details on how mappings and model transformation rules between SEA_ML++ and the target PSMMs are realised as well as how codes are generated from PSMs can be found in [4].

As the last step, the developer needs to add his/her complementary codes, a.k.a. delta codes, to the generated architectural code to have fully functional system. However, some agent development languages, such as JACK, have their graphical editor in which the developer can edit the structure of MAS code. The generated codes achieved from the previous step can be edited and customised to add more platform-specific details which helps to reach more detailed agents and artifacts. Then the delta code can be added to gain the final code.

It is important to note that, although all above-mentioned steps are supported by SEA_ML++ to be done automatically, at any stage the developer may intervene in this development process if he/she wishes to modify or customise the generated agents and artifacts.

8.6. Development of a multi-agent garbage collection CPS

In this section, the design and implementation of a multi-agent garbage collection system, as an agent-based CPS is taken into consideration. This CPS will be executed on the Jason agent platform. Jason provides a platform for the development of BDI agents and MASs. It is a Java-based interpreter for an extended version of a Prolog-like logic programming language for BDI agents, called AgentSpeak.

The following subsections elaborate how the required CPS is designed and implemented according to the MDE process introduced in Section 8.5 and then demonstrate the execution of the modelled multi-agent garbage collection CPS.

8.6.1 System design

In this subsection, we discuss the design of the multi-agent garbage collection CPS. System design is carried out by providing MAS models from different viewpoints of the system using SEA_ML++ language.

The agent-based garbage collection CPS consists of different types and numbers of agents, which cooperate with each other to collect garbage from the environment. The system design phase is performed by considering different viewpoints of the SEA_ML++ language. MAS and Organisation viewpoint provides an overview of the system which is shown in Fig. 8.4. Considering the general structure of the system, the garbage collection CPS constitutes of two organisations, one covering the other, called *MAS_Organisation* which include *GarbageFinder* and *GarbageBurner* agents, and *CollectorOrganisation* where *GarbageCollector* agents reside. It should be noted that the entities given in this overview can be regarded as stereotypes and there may be many instances of these entities in the real system implementation. For example, there may be many agents of the type *GarbageCollector* running on this system, and in fact they are expected to be more than one in a real implementation.

In this multi-agent CPS, a *GarbageFinder* agent handles the interaction between *GarbageCollectors*. This agent is responsible for finding available garbages in the environment and assigning proper *GarbageCollector* agents to collect them. *GarbageFinder* agent interacts with all candidate *GarbageCollector* agents to assign collecting garbage mission, and inform *GarbageBurner* agent about this assignment.

Garbage Collector agents represent the main entities of the designed CPS. These agents receive garbage locations and carry these pieces of garbage to the *GarbageBurner* agent to get rid of them.

The *GarbageBurner* agent is responsible for removing the garbage, which is brought to him.

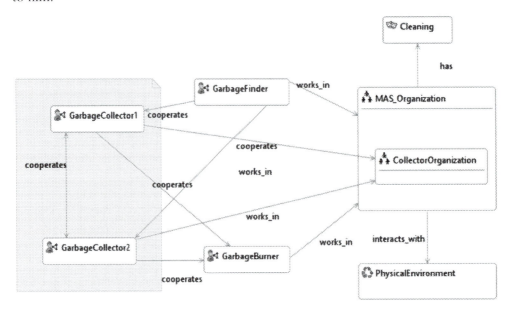

Figure 8.4 Overview of multi-agent garbage collection CPS designed in SEA_ML++ Editor.

Fig. 8.5 shows an instance model of SEA_ML++'s Agent Internal Viewpoint, which demonstrates the general internal process of the *GarbageCollector* agent for collecting garbage.

A *GarbageCollector* agent has only one *Goal* called "CollectTheGarbarges". To accomplish this goal, the agent has two different *Beliefs* and four different *Plans*. Considering the *Beliefs*, the agent uses them to know the position of the garbage that it has to collect and the position of *GarbageBurner* agent, which is responsible for destroying the garbage. Thanks to the four different plans that agent uses to achieve the *Goal*, it can go to the garbage that it knows, and pick and bring it to the GarbageBurner agent to eliminate it. While carrying out these plans, it plays three different roles.

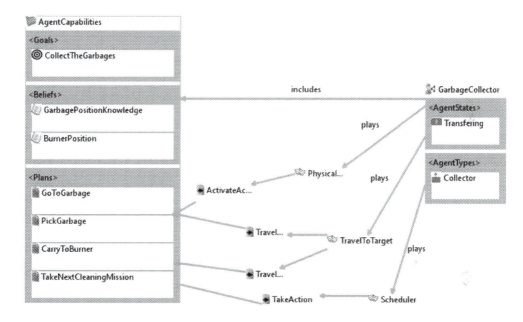

Figure 8.5 Agent internal diagram of the garbage collector agent.

As discussed in Section 8.5, the models provided for the design of the CPS are used in the implementation phase. To this end, the conceptual mappings between SEA_ML++ and Jason platform elements, shown in Table 8.1, are used. Hence, it will be possible to transform SEA_ML++ model instances to Jason platform models and then execute them inside Jason agent execution platform via a series of M2M and M2T transformations discussed in Section 8.5. It is worth indicating that these mappings and transformations are all built-in SEA_ML++ and they are used and executed behind the scene without any human intervention, i.e. CPS developers do not need to deal with them. The implementation is discussed in detail in the next subsection.

Table 8.1 The conceptual mappings between some of the SEA_ML++ and Jason elements.

SEA_ML++ concepts	Jason concepts
MAS/Organisation	MAS
Agent	Agent
Plan	Plan
Behaviour	Actions
Agent state	Belief Base
Agent type	Agent Class
Goal	Goal
Role	Goal
Belief	Belief
Fact	Literal
Environment	Environment

8.6.2 System development

In this study, the proposed multi-agent garbage collection CPS is implemented using the Jason platform. Jason is selected as it is one of the widely accepted Java-based BDI MAS development platform with ongoing support.

Based on the proposed methodology, M2M transformations are applied for transforming the models designed in SEA_ML++ as platform-independent models and JACK BDI agents as platform-specific models. Then the M2T rules are applied on platform-specific models (Jason models) for the generation of ASL files, which include Jason BDI agent codes. As an example of the generated code, see the content of the file called "garbageCollector.asl" shown in Fig. 8.6. This file consists of Prolog-like agent codes. Although the codes generated for MAS can be executed directly in the Java-based interpreter of Jason environment, some of the business logics which are case specific are missing. Therefore, additional codes are added to these generated codes, called delta codes, to have a fully functional system.

The source code generated using SEA_ML++ models includes architectural code. The generated mechanism uses the syntactically correct templates. Also, the models are controlled by the language during the semantic control stage. This prevents many semantic errors in the code when compared with a manual development. So, it is assumed that the code does not have any syntactic errors at this level. However, the delta code must be added manually to create behavioural logic and they may include some syntactic or semantic errors. Manual completion of this code is required for both the MAS and the CPS (Environment) parts of the system. The interested reader may find an extensive discussion of the evaluation of SEA_ML++'s code generation performance and the degree of manual code completion in [7] and [4]. As can be seen in these empirical studies, it is possible to generate more than 80% of a MAS software just by modelling in SEA_ML++.

In Fig. 8.6, the codes given in lines 24–35 were generated and delta codes were added. They create the behavioural logic of the *takeNextCleaningMission* agent plan. When we examine the relevant code snippet, we see that the garbage collector agent searches the location of the garbage on its own belief base and then runs the plan to go to the location of the garbage. Then the garbage collector agent will execute the relevant plans, respectively, to pick the garbage and carry it to the garbage burner agent. Similarly, the other agent plans are generated and some delta codes are added to give the fully functional behavioural logic.

```
1    // Agent garbageCollector in project garbageWord
2
3    /* Initial beliefs and rules */
4
5    /* Initial goals */
6
7    /* Plans */
8
9    +garbage(X,Y)[burnerLocation(X1,Y1),source(Ag)] : not .desire(takeNextCleaningMission) <-
10       +burnerLoc(X1,Y1);
11       .print("Garbage has been detected in ",X," and ",Y," coordinates.");
12       .send(Ag, askOne, garbageAssigned(X,Y),Reply,3000);
13       !makeDecision(Reply,garbage(X,Y)).
14
15
16   +!makeDecision(Reply,garbage(X,Y)): Reply == false <-.print("I will start my next cleaning mission.");
17      !takeNextCleaningMission.
18
19
20   -!makeDecision(Reply,garbage(X,Y)): true <- .print("Abort the cleaning mission"); .abolish(garbage(X,Y)).
21
22   +!takeNextCleaningMission
23      <-
24         ?garbage(Xg,Yg);
25         ?pos(X,Y);
26         -+pos(last,X,Y);
27         !goToGarbage(Xg,Yg);
28         !pickGarbage(Xg,Yg);
29         ?burnerLoc(Xb,Yb);
30         !carryToBurner(Xb,Yb);
31         .print("The cleaning task has been completed.");
32         .print("Move to previous position.")
33         ?pos(last,Xo,Yo);
34         goTo(Xo,Yo);
35         -+pos(Xo,Yo).
36
37   +pickedGarbage(X,Y):true <- delGarbage(X,Y); pickGarbage.
38   -pickedGarbage(X,Y):true <- dropGarbage.
39
40   +!goToGarbage(X,Y):garbage(X,Y)<-.print("going to the garbage in ",X," and",Y," coordinates."); goTo(X,Y);-+pos(X,Y).
41   +!goToGarbage(X,Y).
42
43   +!pickGarbage(X,Y):garbage(X,Y)<-.print("I pick the garbage");
44                    ?burnerLoc(X1,Y1); .abolish(garbage(X,Y));+pickedGarbage(X,Y).
45   +!carryToBurner(X,Y):pickedGarbage(X1,Y1)<-?burnerLoc(Xb,Yb); .print("I carry the garbage");goTo(Xb,Yb); -+pos(X,Y);
46   .send(garbageBurner,achieve,burnGarbage(pickedGarbage(X1,Y1))); -pickedGarbage(X1,Y1).
47
48   +?add_proposal(X): X==false.
49   +?add_proposal(X): not X==false <- .fail.
```

Figure 8.6 The Garbage Collector ASL file.

As another example, the codes in lines 12–19 of Fig. 8.7 were generated and delta codes added, too. These codes form the behavioural logic of the nextGarbage plan. Here, the position of a randomly generated garbage in the environment is sent both to the GarbageBurner agent and to all GarbageCollector agents in the environment.

```
1  // Agent garbageFinder in project garbageWorld
2
3  /* Initial beliefs and rules */
4
5  /* Initial goals */
6
7  !nextGarbage.
8
9  /* Plans */
10
11 +!nextGarbage : true <-
12    .random(R1); X = math.round(20*R1);
13    .random(R2);  Y = math.round((20*R2));
14    .print("I find a garbage located in ",X," and ",Y);
15    +garbage(X,Y);
16    .send(garbageBurner,askOne,pos(Xb,Yb),pos(Xb,Yb));
17    .send([garbageCollector1,garbageCollector2],tell,garbage(X,Y)[burnerLocation(Xb,Yb)]);
18    .wait(1000);
19    !!nextGarbage.
20
21 +garbage(X,Y) : true <- addGarbage(X,Y).
22 -garbage(X,Y) : true <- delGarbage(X,Y).
23
24 +burnedGarbage(X,Y)[assigned(Ag),source(garbageBurner)]<--garbage(X,Y);-garbageAsigned(X,Y,_);
25 .print("Garbage located in ",X," and ",Y," has been cleaned by ",Ag).
26
27 +?garbageAsigned(X,Y)[source(Ag)]: not garbageAsigned(X,Y,_) <- +garbageAsigned(X,Y,Ag).
```

Figure 8.7 The Garbage Finder ASL file.

8.6.3 Demonstration

To demonstrate the implemented system, this section shows the system execution consisting *GarbageCollector*, *GarbageBurner* and *GarbageFinder* agents.

In Fig. 8.8, the user interface of the system when the system is started to run, is shown. In this interface, the user just can see the graphical representation of the agents and the garbage on a grid model.

As soon as the system executes inside the Jason platform, the *GarbageFinder* agent finds the garbage and sends its position to all other agents. In this way, *GarbageCollector* agents contact with the *GarbageFinder* agent to be assigned to the relevant garbage collection task. Then one of them, which is not busy by any other task, is assigned by *GarbageFinder* to collect the relevant garbage. The *GarbageCollector* agent, who takes the task, picks the garbage and carries it to the *GarbageBurner* agent. Thus, *GarbageBurner* agent destroys the garbage. After completing the task, the *GarbageCollector* agent returns to its pre-task position to wait for a new task. The process described above is shown in the screenshots taken from the GUI of the execution environment given in Fig. 8.9, Fig. 8.10 and Fig. 8.11.

An excerpt from the console output of the whole execution of the system is given in Fig. 8.12. As can be seen, all agents of the garbage collection CPS, which are modelled with SEA_ML++, were initialised inside Jason platform and their plans were successfully executed to provide the required agent interactions.

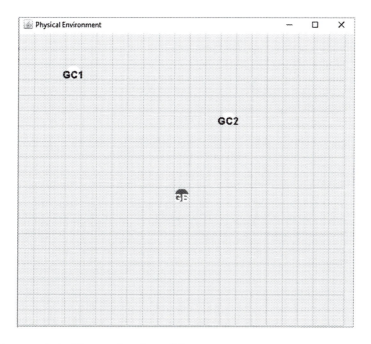

Figure 8.8 A screenshot of the user interface of the system.

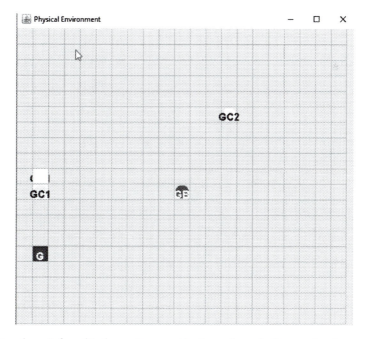

Figure 8.9 A garbage is found in the environment by the garbage finder agent and one of the garbage collector agents goes to pick it up.

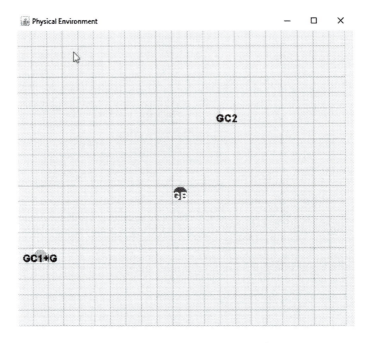

Figure 8.10 The Garbage Collector Agent 1 picks the garbage up.

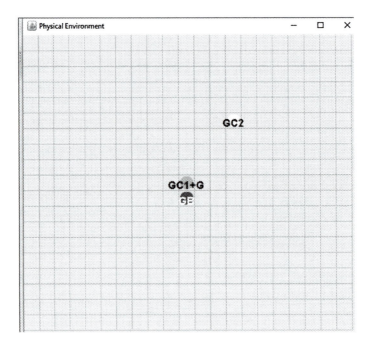

Figure 8.11 The Garbage Collector Agent 1 brings picked garbage to the Garbage Burner Agent.

Figure 8.12 Screenshot of the console output showing the initialisation & interaction of agents.

8.7. Conclusion

In this chapter, we discussed the modelling and development of CPS using agents and MAS. To this end, a generative agent modelling language, called SEA_ML++ was employed. This DSML can support the CPS designer to model different aspects using various viewpoints such as MAS/Organisation, Agent Internal, Plan, Interaction, Environment and so on.

In addition, we introduced an MDE methodology in which SEA_ML++ can be used to design agent-based CPS and implement these systems on various agent execution platforms. To give some flavour of using this methodology, the development of a multi-agent garbage collection CPS was taken into consideration. The conducted study demonstrated how this CPS can be designed according to the various viewpoints of SEA_ML++ and then implemented and executed on Jason BDI agent platform.

Use of SEA_ML++ improved the way of both design and implementation of the agent-based CPS. Based on the conducted case study, one can deduce that a graphical syntax of a MAS DSML facilitates the design of a CPS which is composed of various interacting agents. It is possible to model the structure of an agent-based CPS according to different viewpoints before implementing and deploying the system. A developer can also benefit from the visual modelling environment to check the modelled system at the early phases of the design which may pave to minimise the errors during the implementation. We also showed that the translational semantics of the SEA_ML++ language lead auto-generation of the source code required to execute the agent-based CPS. Hence, the system development process can be shortened with the application of the MDE methodology introduced in this chapter.

In the future, the platform support of SEA_ML++ can be extended with new agent execution environments in addition to Jason. Hence, SEA_ML++ models can be used to implement agent-based CPS on various platforms. However, just creating new transformations to these platforms will not be enough for this purpose. Probably both the syntax and the semantics definitions of the language will also need to be updated and extended when the diversity of main CPS components including sensors, actuators, networks and other physical components are taken into consideration.

Acknowledgements

This study was funded as a bilateral project by the Scientific and Technological Research Council of Turkey (TUBITAK) under grant 115E591 and the Portuguese Foundation for Science and Technology (FCT) under grants FCT/MCTES TUBITAK/0008/2014 and FCT/MCTES PEst UID/CEC/04516/2013. The authors would also like thank the European Cooperation in Science & Technology (COST) Action networking mechanisms and support of COST Action IC1404: Multi-Paradigm Modelling for Cyber-Physical Systems (MPM4CPS). COST is supported by the EU Framework Programme Horizon 2020.

References

[1] S. Russell, P. Norvig, Artificial Intelligence—A Modern Approach, 3rd ed., Pearson, 2016.
[2] M. Wooldridge, N.R. Jennings, Intelligent agents: theory and practice, Knowledge Engineering Review 10 (2) (1995) 115–152.

[3] S. Demirkol, S. Getir, M. Challenger, G. Kardas, Development of an agent based E-barter system, in: 2011 International Symposium on Innovations in Intelligent Systems and Applications, IEEE, 2011, pp. 193–198.

[4] M. Challenger, B.T. Tezel, O.F. Alaca, B. Tekinerdogan, G. Kardas, Development of semantic web-enabled BDI multi-agent systems using SEA_ML: an electronic bartering case study, Applied Sciences 8 (5) (2018) 688.

[5] G. Kardas, M. Challenger, S. Yildirim, A. Yamuc, Design and implementation of a multiagent stock trading system, Software, Practice & Experience 42 (10) (2012) 1247–1273.

[6] E.D. Likotiko, D. Nyambo, J. Mwangoka, Multi-agent based IoT smart waste monitoring and collection architecture, arXiv:1711.03966.

[7] M. Challenger, G. Kardas, B. Tekinerdogan, A systematic approach to evaluating domain-specific modeling language environments for multi-agent systems, Software Quality Journal 24 (3) (2016) 755–795.

[8] V. Mascardi, D. Weyns, A. Ricci, C.B. Earle, A. Casals, M. Challenger, A. Chopra, A. Ciortea, L.A. Dennis, Á.F. Díaz, et al., Engineering multi-agent systems: state of affairs and the road ahead, ACM SIGSOFT Software Engineering Notes 44 (1) (2019) 18–28.

[9] G. Kardas, B.T. Tezel, M. Challenger, Domain-specific modelling language for belief–desire–intention software agents, IET Software 12 (4) (2018) 356–364.

[10] T. Miranda, M. Challenger, B.T. Tezel, O.F. Alaca, A. Barišić, V. Amaral, M. Goulao, G. Kardas, Improving the usability of a MAS DSML, in: International Workshop on Engineering Multi-Agent Systems, in: Lecture Notes in Artificial Intelligence, vol. 11375, Springer, Cham, 2019, pp. 55–75.

[11] J.M. Bradley, E.M. Atkins, Optimization and control of cyber-physical vehicle systems, Sensors 15 (9) (2015) 23020–23049.

[12] R. Fujimoto, C. Bock, W. Chen, E. Page, J.H. Panchal, Research Challenges in Modeling and Simulation for Engineering Complex Systems, Springer, 2017.

[13] M. Brambilla, J. Cabot, M. Wimmer, Model-Driven Software Engineering in Practice, ©Morgan & Claypool Publishers, 2017.

[14] M. Amrani, D. Blouin, R. Heinrich, A. Rensink, H. Vangheluwe, A. Wortmann, Towards a formal specification of multi-paradigm modelling, in: 2019 ACM/IEEE 22nd International Conference on Model Driven Engineering Languages and Systems Companion (MODELS-C), IEEE, 2019, pp. 419–424.

[15] E. Azadi Marand, E. Azadi Marand, M. Challenger, DSML4CP: a domain-specific modeling language for concurrent programming, Computer Languages, Systems and Structures 44 (2015) 319–341.

[16] C. Durmaz, M. Challenger, O. Dagdeviren, G. Kardas, Modelling Contiki-based IoT systems, in: OASIcs—OpenAccess Series in Informatics, vol. 56, Schloss Dagstuhl-Leibniz-Zentrum fuer Informatik, 2017, pp. 5:1–5:13.

[17] S. Arslan, M. Challenger, O. Dagdeviren, Wireless sensor network based fire detection system for libraries, in: Computer Science and Engineering (UBMK), 2017 International Conference on, IEEE, 2017, pp. 271–276.

[18] L. Ozgur, V.K. Akram, M. Challenger, O. Dagdeviren, An IoT based smart thermostat, in: 2018 5th International Conference on Electrical and Electronic Engineering (ICEEE), IEEE, 2018, pp. 252–256.

[19] U. Kulesza, A. Garcia, C. Lucena, P. Alencar, A generative approach for multi-agent system development, in: International Workshop on Software Engineering for Large-Scale Multi-agent Systems, 2004, pp. 52–69.

[20] E.J.T. Gonçalves, M.I. Cortés, G.A.L. Campos, Y.S. Lopes, E.S. Freire, V.T. da Silva, K.S.F. de Oliveira, M.A. de Oliveira, MAS-ML 2.0: supporting the modelling of multi-agent systems with different agent architectures, The Journal of Systems and Software 108 (2015) 77–109.

[21] B.T. Tezel, M. Challenger, G. Kardas, A metamodel for Jason BDI agents, in: 5th Symposium on Languages, Applications and Technologies (SLATE'16), Schloss Dagstuhl-Leibniz-Zentrum fuer Informatik, 2016, pp. 1–9.

[22] C. Hahn, C. Madrigal-Mora, K. Fischer, A platform-independent metamodel for multiagent systems, Autonomous Agents and Multi-Agent Systems 18 (2) (2009) 239–266.

[23] G. Kardas, A. Goknil, O. Dikenelli, N.Y. Topaloglu, Model driven development of semantic web enabled multi-agent systems, International Journal of Cooperative Information Systems 18 (02) (2009) 261–308.

[24] M. Challenger, S. Getir, S. Demirkol, G. Kardas, A domain specific metamodel for semantic web enabled multi-agent systems, in: International Conference on Advanced Information Systems Engineering, Springer, Berlin, Heidelberg, 2011, pp. 177–186.

[25] I. Garcia-Magarino, Towards the integration of the agent-oriented modeling diversity with a powertype-based language, Computer Standards & Interfaces 36 (6) (2014) 941–952.

[26] J.M. Gascueña, E. Navarro, A. Fernández-Caballero, Model-driven engineering techniques for the development of multi-agent systems, Engineering Applications of Artificial Intelligence 25 (1) (2012) 159–173.

[27] S. Demirkol, M. Challenger, S. Getir, T. Kosar, G. Kardas, M. Mernik, A DSL for the development of software agents working within a semantic web environment, Computer Science and Information Systems 10 (4) (2013) 1525–1556.

[28] F. Bergenti, E. Iotti, S. Monica, A. Poggi, Agent-oriented model-driven development for JADE with the JADEL programming language, Computer Languages, Systems and Structures 50 (2017) 142–158.

[29] D. Sredojević, M. Vidaković, M. Ivanović, ALAS: agent-oriented domain-specific language for the development of intelligent distributed non-axiomatic reasoning agents, Enterprise Information Systems 12 (8–9) (2018) 1058–1082.

[30] J. Silva, A. Barišić, V. Amaral, M. Goulão, B.T. Tezel, O.F. Alaca, M. Challenger, G. Kardas, Comparing the usability of two multi-agents systems DSLs: SEA_ML++ and DSML4MAS study design, in: 3rd International Workshop on Human Factors in Modeling (HuFaMo' 18) held under ACM/IEEE 21st International Conference on Model Driven Engineering Languages and Systems (MODELS), 2018, pp. 1–8.

[31] M. Reimann, C. Ruckriegel, S. Mortimer, S. Bageritz, M. Henshaw, C.E. Siemieniuch, M.A. Sinclair, P.J. Palmer, J. Fitzgerald, C. Ingram, et al., Road2CPS priorities and recommendations for research and innovation in cyber-physical systems, © Steinbeis-edition, 2017.

[32] S. Engell, R. Paulen, M.A. Reniers, C. Sonntag, H. Thompson, Core research and innovation areas in cyber-physical systems of systems, in: International Workshop on Design, Modeling, and Evaluation of Cyber Physical Systems, Springer, 2015, pp. 40–55.

[33] H. Thompson, R. Paulen, M. Reniers, C. Sonntag, S. Engell, Analysis of the state-of-the-art and future challenges in cyber-physical systems of systems, EC FP7 project 611115.

[34] T. Bures, D. Weyns, B. Schmerl, J. Fitzgerald, A. Aniculaesei, C. Berger, J. Cambeiro, J. Carlson, S.A. Chowdhury, M. Daun, et al., Software engineering for smart cyber-physical systems (SEsCPS 2018)-workshop report, ACM SIGSOFT Software Engineering Notes 44 (4) (2019) 11–13.

[35] H. Vangheluwe, Multi-paradigm modelling of cyber-physical systems, in: SEsCPS@ ICSE, 2018, p. 1.

[36] D. Morozov, M. Lezoche, H. Panetto, Multi-paradigm modelling of cyber-physical systems, IFAC PapersOnLine 51 (11) (2018) 1385–1390.

[37] F. van den Berg, V. Garousi, B. Tekinerdogan, B.R. Haverkort, Designing cyber-physical systems with aDSL: a domain-specific language and tool support, in: 2018 13th Annual Conference on System of Systems Engineering (SoSE), IEEE, 2018, pp. 225–232.

[38] Y. Hu, X. Zhou, Cps-agent oriented construction and implementation for cyber physical systems, IEEE Access 6 (2018) 57631–57642.

[39] L. Sakurada, J. Barbosa, P. Leitão, G. Alves, A.P. Borges, P. Botelho, Development of agent-based CPS for smart parking systems, in: IECON 2019-45th Annual Conference of the IEEE Industrial Electronics Society, vol. 1, IEEE, 2019, pp. 2964–2969.

[40] D. Calvaresi, M. Marinoni, A. Sturm, M. Schumacher, G. Buttazzo, The challenge of real-time multi-agent systems for enabling IoT and CPS, in: International conference on web intelligence, 2017, pp. 356–364.

[41] Jonas Queiroz, Paulo Leitão, José Barbosa, Eugénio Oliveira, Distributing intelligence among cloud, fog and edge in industrial cyber-physical systems, in: Proceedings of the 16th International Conference on Informatics in Control, Automation and Robotics - vol. 1: ICINCO, INSTICC, SciTePress, 2019, pp. 447–454.

[42] J. Queiroz, P. Leitão, J. Barbosa, E. Oliveira, Agent-based approach for decentralized data analysis in industrial cyber-physical systems, in: International Conference on Industrial Applications of Holonic and Multi-Agent Systems, Springer, 2019, pp. 130–144.

[43] M. Challenger, S. Demirkol, S. Getir, M. Mernik, G. Kardas, T. Kosar, On the use of a domain-specific modeling language in the development of multiagent systems, Engineering Applications of Artificial Intelligence 28 (2014) 111–141.

[44] S. Demirkol, M. Challenger, S. Getir, T. Kosar, G. Kardas, M. Mernik, SEA_L: a domain-specific language for semantic web enabled multi-agent systems, in: 2012 Federated Conference on Computer Science and Information Systems (FedCSIS), IEEE, 2012, pp. 1373–1380.

[45] D. Moody, The "physics" of notations: toward a scientific basis for constructing visual notations in software engineering, IEEE Transactions on Software Engineering 35 (6) (2009) 756–779.

[46] G. Kardas, Z. Demirezen, M. Challenger, Towards a DSML for semantic web enabled multi-agent systems, in: Proceedings of the International Workshop on Formalization of Modeling Languages, 2010, pp. 1–5.

[47] K. Kravari, N. Bassiliades, A survey of agent platforms, Journal of Artificial Societies and Social Simulation 18 (1) (2015) 1–18.

[48] S. Getir, M. Challenger, G. Kardas, The formal semantics of a domain-specific modeling language for semantic web enabled multi-agent systems, International Journal of Cooperative Information Systems 23 (3) (2014) 1–53.

CHAPTER 9

CREST – a DSML for hybrid CPS modelling[☆]

Stefan Klikovits and Didier Buchs
University of Geneva, Carouge, Switzerland

9.1. Introduction

In recent decades, the modelling of any non-trivial of system, including cyber-physical system (CPS) has evolved to a standard systems engineering approach. To this extent, models have been used in various forms and depending on particular influence and timing of the model in the engineering process it is possible to refer to the process using specific terminology such as model-driven development (MDD), model-driven engineering (MDE), and model-based engineering (MBE) [1]. All these subdisciplines share the fact that system models provide important benefits to the development process, as they permit the upfront analysis, testing, simulation and verification. The arguably most important feature of models is their capability to abstract over the "real system" and thereby ignore unimportant features. For example, when studying a lamp's illumination capabilities in a regular home, it is usually possible to ignore factors such as air humidity, temperature, altitude and precise chemical composition of the transport medium. Even though these factors (theoretically) do influence the transport medium (i.e. the air between lamp and illuminated object), the propagation of light and thus perhaps alter the illumination, in most cases this influence is negligibly minuscule.[1] Models of this system should therefore rather focus on the model of the lamp (e.g. its illumination angle) and the position of the object (e.g. its size and distance).

However, this abstraction is also reflected in the type of model that is created. In the most basic classification, it is possible to distinguish systems where system modifications are observed at discrete points in time and systems where changes are continuous. In the former, changes are usually observed as *events* that occur at particular points in time. Discrete event models are often created for systems with a clearly identifiable set of

---

[☆] This project is supported by FNRS STRATOS: Strategy based Term Rewriting for Analysis and Testing Of Software, the Hasler Foundation, 1604 CPS-Move and COST IC1404: MPM4CPS.

[1] Nevertheless, we can imagine extreme situations where such factors can play an important role and significantly influence the measured result. Examples can be found for example in high-energy particle physics, where "Ring-imaging Cherenkov detectors" (RICH) are used to measure subatomic particles by analysing photon speeds within a special gas mixture.

Multi-Paradigm Modelling Approaches for Cyber-Physical Systems
https://doi.org/10.1016/B978-0-12-819105-7.00014-3

system states and transitions between them. For example, the functionality of many electrical devices, machines and software systems can be intuitively expressed in this fashion. When looking at an electrical fan, it usually has several pre-defined settings (e.g. `Off`, `Speed-1`, `Speed-2`) which define the rotation speed of the ventilator. When a user turns the switch to another state, her action triggers a discrete event and the system reacts to the new switch setting. The system's behaviour remains again stable, until another event is observed and the system switches again to another state.

On the other hand, most models of natural phenomena and physical processes represent continuous changes of a system. For example, when putting a pot of water onto a cooking stove, the temperature of the water gradually increases until the water boils and evaporates. Such systems have to be defined using ordinary differential equations (ODEs) or partial differential equations (PDEs), as it is generally impossible to define discrete states and transitions that represent the system state.

While continuous models are well-suited to represent physical processes, and discrete systems can be used to capture the behaviour of digital systems, the rise of CPS created the need for the expression of both concepts within the same system. For example, when modelling the use of an electrical heater within an office, the heater's functionality can be modelled as discrete-state machine. The heater's output, however, does not switch discretely, but gradually increases in state `On` and decreases in state `Off`. This is due to the fact that the heater's internal heating rod takes time to heat up and cool down. Thus, in each discrete state of the system, there is the need of a continuous function (e.g. an ODE) that calculates the output.

Such systems, whose behaviour description requires discrete and continuous concepts are commonly referred to as *hybrid*. In this chapter we will analyse the current state of hybrid systems modelling, and introduce various formalisms that have been developed to support hybrid systems. In the second part, we introduce CREST, a domain-specific modelling language (DSML) for the modelling, simulation and verification of hybrid systems.

9.2. Hybrid formalisms

The need for the combination of discrete event systems (DEVS) and continuous (i.e. differential equation) systems has already been known for a long time and described in various publications (e.g. [2–4]) and books (e.g. [5]). Especially the latter analyses these multi-formalism approach and the merging of discrete events, discrete time and differential equation systems.

One of the most generic formalisms for the modelling of hybrid systems are hybrid automata (HAs) [6]. HAs are extensions of regular classic automata (e.g. finite-state machines (FSMs) [7]) that model a set of continuous variables. In each of the automaton's states, ODEs are used to describe the variables' evolution. Transitions are annotated with

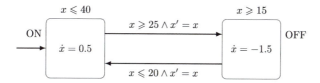

Figure 9.1 Example of a hybrid automaton that models a very simple heater. x represents the temperature value, $\dot{x}$ the rate at which x changes and x' is the value that x takes after the transition.

jump expressions which serve as guard conditions and discrete value changes. Guards restrict when a condition can be taken and the value changes allow the discrete update of variable values through transitions (e.g. double the value).

Fig. 9.1 shows an example of a HA using the usual notation form. The system models the behaviour of a simplified heater. It consists of two states (called *locations*), connected by two transitions. The entire system uses one continuous variable x that represents the temperature of the room that the heater is placed in. If the device is in location ON, the room temperature x grows by 0.5 degrees centigrade, when it is OFF it drops by 1.5 degrees. Commonly, these change rates are defined as "dotted" variables (e.g. $\dot{x}$) and written inside the respective locations. Additionally, locations are annotated with invariants that limit the time that an automaton can spend in a state. In the example, we see that the heater can only remain ON until the room temperature reaches 40 degrees. At this point the automaton has to take a transition. As stated above, transitions are annotated with jump expressions. The jump expression for the transition to OFF for example states that it can only be taken when the temperature is at least 25 degrees ($x \geqslant 25$) and that the temperature value is not changed by the transition (i.e. $x' = x$).

Semantics – May and Must transitions

The way in which HAs are specified allows for two different semantics that severely impact the model's behaviour. The more common *may* semantics expresses that a HA can remain in any location as long as all location invariants are satisfied. Any enabled transition may, but does not necessarily have to be taken. In the above example, we see that the system can remain in location ON until x reaches 40. However, the transition to OFF already becomes enabled when $x = 25$. The result is that there are infinitely many system configurations at which the transition may be taken, as any $20 \leqslant x \leqslant 40$ is a configuration.

It becomes evident that this semantics is highly powerful but also introduces a lot of complexity which renders many verification problems (e.g. reachability, liveness) undecidable [8]. Some subclasses, however, such as linear and rectangular HAs, are analysable and the boundaries between their decidability and undecidability have been eagerly studied [9–11].

On the other hand, some HA models and tools offer the use of the alternative *must* semantics [12]. These state that a transition has to be taken, as soon as it becomes enabled. In the example, the heater has to switch to OFF as soon as the room temperature reaches or exceeds 20 degrees. Evidently, this semantics and the *must*-constraint simplify the system and reduce the verification task, as the number of possible system states at which the system switches location is reduced to one.

Timed automata

A well-studied subclass of HAs is the timed automaton (TA) [13,14] formalism. A TA is a HA, in which all continuous variables c (called *clocks*) have a rate of $\dot{c} = 1$ and are initialised with value 0. Additionally, a transition can either leave the clock value unchanged ($c' = c$), or reset the clock's value to zero ($c' = 0$). This simplification renders TA decidable for reachability problems [15].

Various extensions have been proposed that slightly increase the TA's capabilities, without rendering it as generic as HA. To name a few, we might look at stopwatch automata [16], where clock advances can be paused (i.e. $\dot{c} = 0 \vee \dot{c} = 1$), interrupt automata [17], which define different interrupt levels and allow only one active clock per level and hourglass automata [18], where clocks have maximum values and can run backwards. There are also techniques describing TA with independently evolving clocks [19]. These modifications obviously also influence its language properties and verification complexity.

9.2.1 Timed and hybrid automata tools

The popularity of TA and hybrid systems (HSs) and their powerful expressiveness led to the development of various tool and languages which allow the specification, analysis, simulation and verification of continuous-time-discrete-state systems. Well-known representatives are UPPAAL [20] and Kronos [21] for TA and Simulink/Stateflow [22], Modelica [23], and HyVisual [24] for HS.

One of the most commonly used ones is Simulink. Simulink is a graphical programming environment for the modelling and analysis of dynamic systems. The tool is extended by Stateflow, a plugin that allows the definition of discrete automata based on non-deterministic Harel statecharts [25]. In combination, Simulink/Stateflow can be used to model HAs. The tool's language allows reactive and parallel modelling and allows for system compositions with data encapsulation.

HyVisual is a HAs application based on the Ptolemy II multi-formalism simulation and verification platform. It allows the definition of HAs with causal influences. It also convinces through its formal semantics, which can be leveraged for simulation and verification. The exclusively graphical modelling environment, however, is difficult to understand for newcomers, while at the same time being a tedious burden for advanced power-users, as pointed out in [26].

Modelica is a textual language specification for the specification of multi-domain systems. It supports the non-causal definition of complex systems. Modelica is purely a language specification and does not offer any reference implementation or formal semantics. It relies on other entities to create implementations of the language. This resulted in the creation of a range of different tools (e.g. OpenModelica [27], Dymola [28]) that support and implement it. A thorough comparison and discussion of various popular HAs tools for simulation and verification is available in [26].

9.2.2 Hybrid extensions of discrete formalisms

The need for continuous behaviour also led to the adaptation of other existing, discrete formalisms to offer such capabilities. The Discrete Event System Specification (DEVS) [29] for example has been extended to *hybrid DEVS*, which allows continuous evolution of state variables. The introduction of this feature, however, has been shown to have severe drawbacks and renders DEVS more complex. For example, one way is to use discretisation of the continuous behaviour using the *quantised state system* (QSS) [30,31] approach. Alternatively, another solution is to use external data structures for the calculation of continuous behaviour provided to the encapsulated DEVS model. This latter solution, however, needs additionally an adaptation of the DEVS simulation algorithms, as outlined in [32]. Tool support for the former QSS approach has been implemented in PowerDEVS [33], a graphical tool for the modelling of DEVS and QSS systems, the latter allowing for seemingly continuous evolution.

Petri nets [34] are a family of formalisms which allow the efficient representation of concurrent processes within complex systems. In the Petri nets domain, there is a long history of encoding and modifying complex data structures using so-called *high-level Petri nets* (HLPN). In HLPN, transitions are extended with *guards*, which evaluate values and assert a condition before transition firings. This led to the introduction and adaptation of Petri nets to support the notion of time. Time has been added in several forms [35] such as Time Petri nets, where transition firings are restricted to certain time points, Timed Petri nets, where time is a token parameter that can be evaluated in transition guards, and Petri nets with Time Windows, where transitions can only fire within certain time frames.

Petri nets have also been extended to continuous Petri nets [36,37], where a transition consumes and produces infinitesimal amounts tokens. Thus, the actual behaviour can only be evaluated when observing the net over time and the transitions are actually seen as streams of tokens exiting and entering places at given rates. The merging of discrete Petri nets with time and continuous Petri nets results in the creation of hybrid Petri nets [38,39]. This formalism is highly intuitive and comparable to hybrid automaton. In fact, there exist translations between these formalisms, such that the simulation and verification of hybrid Petri nets can benefit from existing HA-tools [40].

9.3. Domain-specific, hybrid modelling using CREST

The large variety of systems and the numerous existing formalisms led to the adaptation of many other formalisms and languages to support hybrid concepts. Zélus [41] for example merges continuous variable evolution with the synchronous language Lustre [42], and AADL's Behavioural Annex [43] introduced automaton-based behaviour to architecture description languages.

In general, systems modelling has been mainly used by creators of large-scale and complex systems. The stakeholders of "classical" CPS domains such as aviation, transport and heavy industry have recognised the potential of these modelling languages and tools early on and use them successfully to control, simulate and verify their systems. While their approaches provide significant benefits to financially potent institutions, creators of small and custom systems, such as home automation and Internet-of-Things systems, often lack the knowledge and resources to use such tools.

In the rest of this chapter, we will discuss the modelling of hybrid systems using CREST [56]. CREST is a DSML that has been developed particularly for the use in small-scale CPS, such as smart homes, office and building automation, automatic gardening systems and similar. The language's target audience consists of novice modellers and non-expert users in the modelling and simulation (M&S) domain, who want to take advantage of the values added by modelling and verification.

An initial analysis [56] reveals that such systems exhibit predominantly three types of behaviour, namely a) the continuous flow of physical resources (e.g. light, heat, electricity) and data signals (e.g. on/off switches, control commands) within a system; b) the state-based behaviour of CPS components and devices; and c) the evolution of the global system states over time.

A more detailed evaluation of representative case study systems led to the discovery of six key aspects that should be supported by a hybrid modelling language, so that these systems can be effectively modelled [45]:

1. *Reactivity*. The goal of CPS is to model components and systems that react to changes in their environment. For example, when the sun sets, a home automation system should adapt and provide another light source.

2. *Parallelism*. While sequential execution is an option in computer-based systems, physical processes advance in parallel and thus require appropriate means to be represented. For example, a tripped electrical fuse will shut down all electrical appliances at the same time.

3. *Synchronism*. Even though most changes in CPS are visible over time, their effects are immediate. For example, a room is (virtually) immediately illuminated by a lamp. The actual time delay is negligible for our target applications. Even for energy saving lamps, whose luminosity increases over time, the transition to the on-state and dissipation of light starts immediately. The synchronism aspect enforces that the modelled systems are synchronised and checked as soon as a system value changes,

the entire system is synchronised and checked for possible changed influences between components.

4. *Locality.* Despite the exchange of data and resources, CPS components usually have states and data that should remain local. For example we can think of an energy saving lamp. Its state, life-time and power consumption are local attributes, independent of other components. Interaction occurs through a well–defined interface, i.e. the power plug and switch.

5. *Continuous Time.* Most CPS deal in some way with timing aspects. Plants require a certain amount of light per day, electricity consumption is measured over time, etc. The chosen formalism should allow the modelling of continuous time and thereby arbitrarily fine (coarse) time steps. The time concept also has to support continuous influences between components (e.g. a pump filling a water tank).

6. *Non-determinism.* When it comes to real-world applications, the evolution of a system is not always predictable. For example, a light bulb might break due to power spikes that occur seemingly random. It should be possible to model a scenario where the next state is unknown.

CREST was developed to satisfy the above requirements, while preserving simplicity, usability and suitability for the target domain. The resulting, intuitive graphical language is easy to grasp, but expressive enough to support complex modelling and verification tasks. CREST builds upon the existing state of the art in CPS modelling and combines features from various modelling paradigms and formalisms. To this extent, a reader familiar with this matter will recognise features of hybrid automata, architecture description languages and synchronous languages. Additionally, CREST offers a formal syntax and semantics that allow for well–defined simulation and verification using well-known techniques such as model checking.

The rest of this section will introduce CREST's graphical syntax and semantics in an informal manner. The interested reader might find the formal structure and an extended discussion in [56].

9.3.1 CREST syntax

CREST's graphical syntax, called *CREST diagram*, aims to facilitate the legibility of architecture and behaviour within the system. The rest of this section uses the concrete example of a growing lamp to introduce the individual CREST concepts. The growing lamp is a device that is used for cultivation of plants. When turned on, it consumes electricity to produce light and heat. The light module is always active when the lamp is turned on using the primary switch. The heat module can be (de-)activated using an additional switch. Fig. 9.2 displays the complete CREST diagram of the growing lamp, subsequently.

CREST models components and systems as **entities**. Following the *locality* aspect, CREST entities define their scope (visually depicted by a black border). The scope

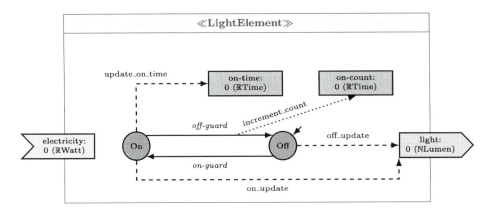

Figure 9.2 A light element that can be used in a growing lamp entity. Note that this diagram annotates updates with the function names, but omits their implementation. These implementations are usually provided as mathematical description or source code implementations alongside the diagram.

limits the information that is used by the component itself and creates cohesion. The entity's communication interface is drawn on the edge of this scope, while the internal structure and behaviour are placed on the inside.

Data is represented in CREST using *ports* that are further used to model the resource within the system. There are three types of ports: **inputs** ⊐, **outputs** ⊃ and **locals** ⊐. Each port is annotated with one specific resource that it can represent. **Resources** are value types consisting of a value domain and a unit. For example the growing lamp specifies units such as Watt or Lumen. Domains are sets of possible individual values, e.g. the natural numbers $\mathbb{N}$, rationals $\mathbb{R}$ or a set of discrete values such as {on, off} (e.g. to define a Switch). Additionally to the type of resource, each port shows its initial resource value (or current value during simulation).

Inputs and outputs make up the entity's communication interface and are thus drawn on the edge of the entity scope, while locals are strictly inside the entity and cannot be accessed from the outside. For example, in Fig. 9.3 the Switch is an input, while on-time and on-count are internal ports.

Discrete behaviour

Each CREST entity defines a FSM using a set of **states** and guarded **transitions** to specify its behaviour. A transition defines a possible discrete advance from one state to another (e.g.). Transition guard functions analyse an entity's and its direct subentities' ports and return a Boolean value that describes whether the automaton can advance to a different state or not. CREST's transition semantics state that a transition has to be taken if it is enabled. This is comparable to the *must*-semantics of HA, as introduced above. Note that, for legibility reasons, complex transition guards are not written directly into the CREST diagram. Instead the guards are annotated with a function name that is executed at runtime. In this case, the detailed functionality is provided alongside the diagram as source code listing or mathematical formula.

Continuous behaviour

Continuous behaviour in CREST is expressed using **updates** (- ➤). Updates are responsible for the continuous modification of an entity's port values. Each update is defined for a specific state and target port that is updated. Similar to guards, their functionality is usually also defined as code or mathematical formula and provided alongside the CREST diagram. The diagram itself identifies the respective update functions using their names. Theoretically, there is one update for each port and state that continuously updates the port's value when the entity's automaton is in the respective state. To avoid redundant information, updates that do not alter the port's value are omitted. The light element for example defines the `update_on_time`, `off_update`. The updates are responsible for the *continuous* aspect of CREST. They can be related to the continuous variable evolution in HAs. However, rather than specifying the rate of continuous variables using ODEs, updates calculate port values based on other ports' values and the amount of time that passes. Thus, when a port value changes, the system has to be tested whether any other ports have to be updated too. This feature therefore serves the *synchronism* requirement that was defined above. CREST's formal semantics assert that dependencies between updates are taken into account and no cyclic definitions are created.

In many CPS there are resources that continuously flow from one port to another, or whose value has to be propagated irrespective of the entity's state. CREST therefore offers the notion of **influences** (—➤), which define relations between exactly two ports. The light element does not specify any influence, but an example can be found in the growing lamp system introduced below. The growing lamp's `fahrenheit_to_celsius` influence for instance is responsible for continuously reading the `room-temperature` input and writing it to the `adder`'s `temp-in` value. Additionally, the influence can optionally modify the source's value before writing it. In this case, `fahrenheit_to_celsius` converts the value to degrees Celsius to match the `adder`'s input port's resource specification.

Finally, CREST defines a third type of resource flow: **actions** (···➤). Actions are linked to transitions and can modify port values when the transition is triggered. They can be compared to value updates during discrete transitions in HAs. As actions are instantaneous they are—similar to influences—also time-independent. The light module defines one action (`increment_count`) that is executed when the transition from `Off` to `On` is triggered. It counts the number of times the lamp has been switched on. Note that influences and actions are in fact syntactic sugar and can be expressed using additional states and updates. However, the language adds them to increase usability.

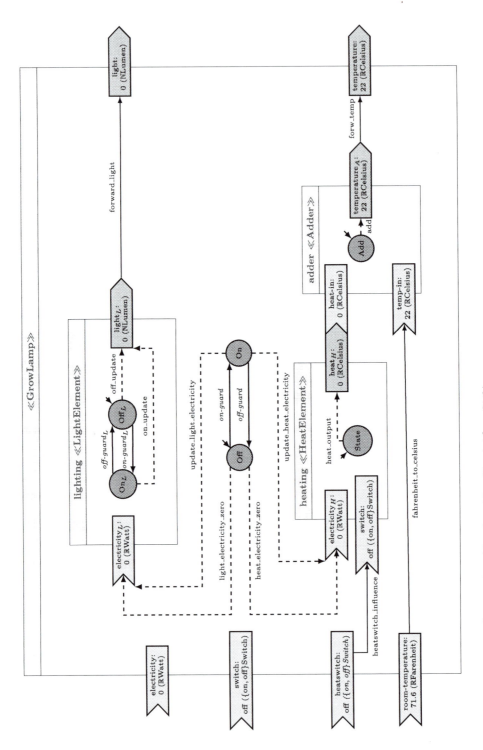

Figure 9.3 A growing lamp entity with subentities. (Reprint from [44].)

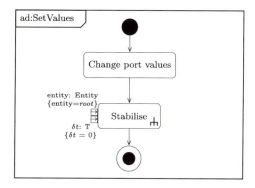

Figure 9.4 The process of setting values. (Reprint from [44].)

Entity composition

The composition of CREST systems follows a strict hierarchy concept, such that there is one single *root* entity. All other entities are placed as subentities in a nested tree-structure inside the *root* entity. Fig. 9.3 shows the example of the growing lamp (GrowLamp) that embeds a (simplified version of the) LightElement[2] and the HeatElement as subentities. This strict hierarchy strengthens the localised entity view, as each component encapsulates its internal subentity structure and therefore can be seen as *black box* from the outside. This facilitates composition, since an entity can be treated as coherent instance, disregarding the inside.

9.3.2 CREST semantics

Similar to most modelling languages, CREST allows two forms of interactions: The setting of input values and the advance of time. After each interaction, the system state has to be tested for transitions that might have become enabled or port values that have to be propagated. This workflow has been schematically depicted as UML 2.5 Activity Diagram in Fig. 9.4. A state with such pending actions is commonly referred to as "unstable", the counter-action as "stabilisation". Stabilisation is an essential concept within CREST. The semantics of CREST's value propagation throughout the hierarchy structure are similar to the semantics of the Functional Mockup Interface's master algorithm [46]. The system's root entity manages the value propagation to and between its direct children. This process is recursive.

---

[2] Note that, for the GrowLamp, the light module's ports and transition guards have been renamed to avoid ambiguities.

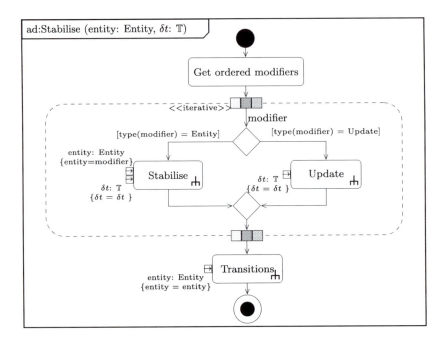

Figure 9.5 The stabilisation process conceptualised as activity diagram. (Reprint from [44].)

Setting values

In the `GrowLamp` example it is visible that a modification of the `electricity` value invalidates the system state. To counteract, this new value has to be propagated to the corresponding inputs of the light and heat modules' input ports. These modules will in turn modify their respective output port values, which will then trigger further propagation. A change of one input value therefore has to be recursively propagated throughout the entire entity hierarchy, starting at the root entity, whose inputs have been modified.

In general, influences, updates and subentities all have the potential to read one port value and wrote another. Collectively, these three are referred to as "modifiers". The above example shows that in CREST modifiers can influence one another, so that the execution of one writes to a port that is read by another modifier. Therefore, the stabilisation of an entity (see Fig. 9.5) has to take these dependencies into account and create an ordered list of modifiers, such that any modifier that reads one value is executed after all modifiers that write to that value. For updates, the process calculates the new port value and then writes it to the specified port.[3] If the modifier is an entity, CREST recursively stabilises this subentity, before continuing execution of the ordered list. This execution supports the localised black box view, as any entity can always assume

---

[3] The activity diagram for the Update process has been omitted due to its simplicity.

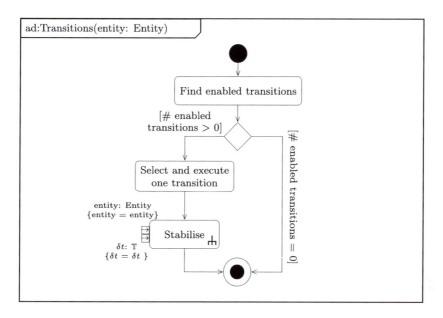

Figure 9.6 The activity diagram showing the transition process. (Reprint from [44].)

that its subentities are stabilised in the right order. Further, the testing for changed input values, and recalculation only upon their change, enforces the *reactivity* principle that we specified as a requirement. The creation of such execution orders is related to Kahn process networks and synchronous language such as Esterel and Lustre, which operate similarly.

After execution of all modifiers, the entity's FSM is searched for enabled transitions from the current state. If a transition is enabled, CREST executes it. Subsequently another stabilisation is started to trigger all updates that are related to the new FSM state. This stabilisation phase will, again, look for enabled transitions and trigger one if applicable. This process is shown in Fig. 9.6.

Advancing time

The stabilisation after time advances (see `ad:Advance`) differs slightly from the one after the modification of port values. In CREST's semantics, this is expressed using the δt parameter. δt states how much time has passed since an update has been last executed and can be used to model continuous-time advances. In fact, the stabilisation process after setting of values, also uses this parameter, except it is set to $\delta t = 0$.

In theory, time advances should be done in infinitesimal timesteps to assert all ports are always up to date. Practically, it is only necessary to assert that all port values are updated before a transition is fired. Thus, it is necessary to discover the next transition

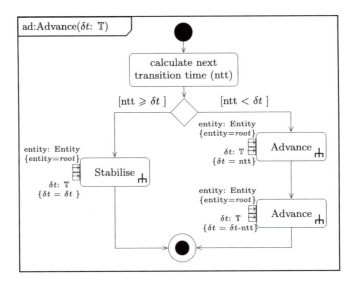

Figure 9.7 The activity diagram showing the two possibilities in the advance process. (Reprint from [44].)

time *ntt*, which signifies how much time has to pass until any transition in the system becomes enabled. CREST's formal semantics does not prescribe the precise functionality for the discovery of *ntt*. The tool implementation of CREST (see next section below) uses an approach that analyses all modifier dependencies within a system that could affect the outcome of the transition guards. It then composes a set of constraints that represents this functionality and uses theorem provers and satisfiability modulo theories (SMT) solvers to find solutions. The minimal solution to the constraint set corresponds to the next transition time *ntt*.[4]

As shown in Fig. 9.7, depending on the value of *ntt* two scenarios are possible:

1. $ntt \geq \delta t$. CREST advances δt and triggers the stabilisation so that all modifiers are executed and all port values updated.
2. $ntt < \delta t$. CREST divides the advance into two steps: In the first phase, the system advances *ntt* time units, i.e. advances until a transition is enabled. The updates and transitions are triggered, followed by stabilisation. Next, CREST recurses on the remaining time $\delta t - ntt$.

These time semantics allow the simulation and verification, and integrate arbitrarily fine (coarse) time advances, while asserting that no "interesting" point in time is missed. This is essential for the precise simulation of cyber-physical systems without the need for an artificial base-clock. The time-based enabling of transitions adds *continuous behaviour*

---

[4] Alternatively, other approaches could be employed, such as search-based algorithms or numerical solvers, which are common in most modern model simulation tools (Simulink, Ptolemy II, etc.)

to the otherwise purely reactive system. In comparison, other synchronous languages such as Lustre rely on external clocks to provide timing signals (ticks) at which the system is synchronised.

9.3.3 Verification

The verification of HAs is non-trivial and, depending on the exact type of HA and the model itself, highly complex and oftentimes even unfeasible. The use of *must*-semantics, as is the case for CREST, reduces this complexity slightly, but still leads to systems that require lots of computation resources for verification tasks.

One very common formal verification method is model checking. In this discipline, the system converted to a set of reachable system states. The verification of system properties (e.g. reachability, liveness) is then performed by searching this so-called *state space* for individual (or sets of) states that fulfil these properties. The complexity arises, as the state spaces of non-trivial systems often contain a very large number of states (often 10^{30} and more) that are difficult to enumerate, store in computer memory and analyse.

Model checking evolved into a de-facto standard, where oftentimes the state space is encoded within a Kripke structure [47]—a form of transition automaton where each state is annotated with system properties—and the formulas that are tested are encoded using [48] and Linear Temporal Logic (LTL) [49] formulas.

In the case of HA and continuous-time systems such as CREST models, however, one problem is the encoding of time information within these structures. Therefore both, Kripke structures and the logic formulas have been extended such that their timed counterparts (Timed Kripke structures and Timed Computation Tree Logic (CTL) (TCTL)) can be used for verification of HA.

This means that TCTL allows the specification of precise information about the point in time when an event can occur. For example, the following phrase can not be specified in CTL, but requires TCTL: *In any case, my alarm will ring until I turn it off within the next hour.*

$$A(\text{alarm rings}) \ U_{\leqslant 60 \ \text{min}} \ (\text{turn off})$$

TCTL's extension is the specification of the "$\leqslant 60$ min"-interval that signifies when the *turn off* event has to be encountered.

An elaborate discussion of TCTL model checking on HA exceeds the scope of this chapter. The interested reader is referred to dedicated publications such as [50] and [51].

9.4. Implementation

Despite the ease of understanding and expressive power of CREST diagrams, CREST is implemented in a classical programming language. In fact, CREST is—similar to

SystemC [52]—developed and distributed as *internal* DSML and uses Python as a host language.

Python was chosen as a target language for three reasons:

1. Python's distribution and package installation allow easy access and extension. It also comes pre-installed on most modern operating systems.
2. It is easy to learn, flexible, has many useful libraries, and a large community.
3. As an interpreted language, Python provides native reflection mechanisms for the modification of class instantiation procedures and dynamic modification of functionality. This enables the hiding of CREST specifics from users, while still enabling the use of the default Python runtime.

As mentioned, CREST is developed and distributed as a set of Python libraries that can be included into any standard Python program. Nevertheless, there are APIs for the use Python's modern browser-based Jupyter[5] runtime, which allows for more interactive development and execution. Some of the integrated features include native plotting of CREST diagrams based on the code and integration of data analysis libraries such as NumPy and Pandas.[6] In the following a small sample system is provided that showcases the use of CREST's Python implementation for the modelling of the growing lamp's light element. The Python listing is provided in Listing 9.1.

Listing 9.1: The definition of the LightElement entity.

```
import crestdsl.model as crest

# use CREST's domain types to specify the domain
watt = crest.Resource(unit="Watt", domain=crest.REAL)
lumen = crest.Resource(unit="Lumen", domain=crest.INTEGER)

class LightElement(crest.Entity):
    # port definitions with resources and an initial value
    electricity = crest.Input(resource=watt, value=0)
    light = crest.Output(resource=lumen, value=0)

    # automaton states - specify one as the current (initial)
    state
    on = crest.State()
    off = current = crest.State()

    # transitions and guards (as lambdas)
    off_to_on = crest.Transition(source=off, target=on,
        guard=(lambda self: self.electricity.value >= 100))
    on_to_off = crest.Transition(source=on, target=off, \
        guard=(lambda self: self.electricity.value < 100))
```

---

[5] https://jupyter.org/.

[6] https://www.numpy.org/; https://pandas.pydata.org.

```
21
22        # updates are annotations
23        @crest.update(state=on, target=light)
24        def set_light_on(self, dt=0):
25            return 800
26
27        @crest.update(state=off, target=light)
28        def set_light_off(self, dt=0):
29            return 0
30
31   my_lamp_obj = LightElement()
```

As the listing shows, the emphasis of CREST's model specification lies on simplicity. Entities are defined as a regular Python classes that inherit from the library's Entity. Furthermore, there is a class provided for each individual model concept (Input, State, Update, etc.) as well as implementations of Python's method decorators to further increase usability (e.g. @influence, @transition).

All entity ports are specified as the class's attributes or decorated methods. CREST's library implementation takes care of the correct instantiation and propagation of information during instantiation of objects. Particular attention was paid to maximise the compatibility with Python's natural development best practices. Therefore, it is possible to inherit from previously defined Entity classes and use __init__-constructors as in any other Python program.

CREST's simulation and verification libraries are created in a similar fashion. For example, Listing 9.2 shows an example of the usage of the Simulator class. Similarly, for models that are time-dependent, the simulator offers an advance(dt) method.

Listing 9.2: An example listing that uses the Simulator library.

```
1   from crestdsl.simulator.simulator import Simulator
2
3   sim = Simulator(my_lamp_obj)
4   sim.stabilise()
5   sim.plot() # shows default behaviour
6
7   my_lamp_obj.electricity.value = 100   # add power
8   sim.stabilise()  # stabilise
9   sim.plot() # state is ON and there is light
```

Fig. 9.8 displays the output of CREST's simulator after the execution of the code above. The simulators plot method creates an interactive CREST diagram that can be analysed and explored. Note how in the first diagram (above), the automaton is in state off, but after setting the electricity input port's value to 100 watts and a stabilisation of the system, the system switches to state on and produces light at its output (diagram below).

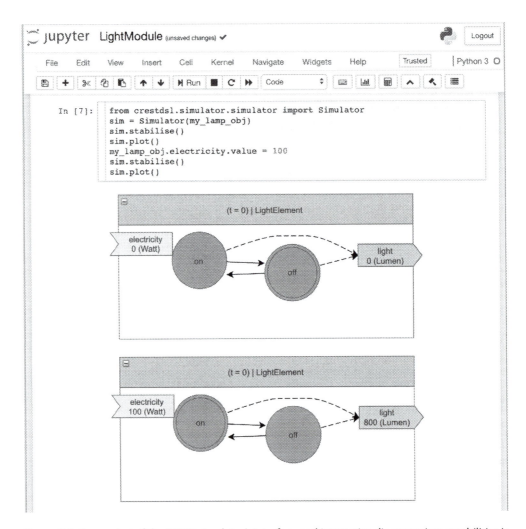

Figure 9.8 Screenshot of the CREST simulator's interface and interactive diagramming capabilities in the Jupyter runtime.

9.5. Discussion

The CREST formalism and DSML provide a simple means to modelling resource flows in small CPS. While the project is still under development and a formal usability evaluation is pending, it is evident that the language facilitates the creation of system models. Our confidence is founded on user reports throughout the language development. Recently, we invited two undergraduate students to use the language without providing them any formal training. Our ambition was to (informally) evaluate the intuitiveness of the development approach.

The students were tasked with various work projects surrounding the project, such as the development of system models (e.g. a smart home setup and an office automation system) and the extension of the DSML's user interface library. Even though the students were not working in a formal experiment setting, we can report positive results. The first (engineering) student successfully implemented all three case study systems based on a system description and the online documentation over the course of three months. He managed to create complex component structures with interdependent entities and even produce solutions for intricate engineering problems such as a linearised model of the heating process within a hot water boiler. He also developed an entity library of parametrisable models for the reuse in future projects. The second (computer science) student acquired knowledge of the structure of CREST models and helped us improve the graphical output and tracing capabilities of CREST. He showed that in a time span of eight weeks, it is possible for a programming-versed developer to become a proficient CREST user and develop domain-specific language (DSL) extensions that can be reintegrated into the library.

Despite this reassuring feedback, a more formal evaluation is required. Thus, we are in the process of developing a DSL usability study based on the principles found in work such as [53] and [54].

We also show that CREST can successfully be used as a basis for further research, for instance as a target formalism for machine learning applications. The DSML's foundation as an embedding within Python facilitates the interoperability and allows the use for automatic calculation of influence functions, as described in [55].

9.6. Summary

This chapter reviews the domain of hybrid systems modelling. Hybrid systems merge continuous evolution of system values such as in natural processes with discrete, usually state-based behaviour descriptions which are very common in electrical and digital devices. Especially for the modelling of cyber–physical system (CPS), the use of hybrid systems is often unavoidable due to the mixture of these inherently opposing concepts. One of the most common formalisms for CPS modelling are hybrid automata, which combine finite-state machines with continuous variables whose evolution is defined by differential equations, but there also exist other formalisms that have been extended to support continuous variable evolutions such as quantised state systems and hybrid Petri nets.

The practical part of this chapter introduces CREST, a hybrid modelling language. CREST is graphical domain-specific modelling language that focuses on the efficient representation of flows of physical resources within a small-scale CPS such as automated gardening applications and smart homes. The language's syntax, semantics and verification concepts are outlined using practical examples. Finally, we present CREST's implementation into the Python host language and present a highly pragmatic internal DSL-based approach to developing modelling languages and tools.

References

[1] M. Brambilla, J. Cabot, M. Wimmer, Model-Driven Software Engineering in Practice, 1st edition, Morgan & Claypool Publishers, 2012.

[2] H. Witsenhausen, A class of hybrid-state continuous-time dynamic systems, IEEE Transactions on Automatic Control 11 (2) (1966) 161–167, doi:10/bc93wv.

[3] L. Tavernini, Differential automata and their discrete simulators, Nonlinear Analysis: Theory, Methods & Applications 11 (6) (1987) 665–683, doi:10/csqvmt.

[4] M. Branicky, V. Borkar, S. Mitter, A unified framework for hybrid control: model and optimal control theory, IEEE Transactions on Automatic Control 43 (1) (Jan. 1998) 31–45, doi:10/c64n32.

[5] B.P. Zeigler, T.G. Kim, H. Praehofer, Theory of Modeling and Simulation, 2nd edition, Academic Press, Inc., Orlando, FL, USA, 2000.

[6] J.-F. Raskin, An introduction to hybrid automata, in: Handbook of Networked and Embedded Control Systems, Springer, 2005, pp. 491–517.

[7] A. Gill, Introduction to the Theory of Finite-State Machines, McGraw-Hill Electronic Sciences Series, McGraw-Hill, 1962.

[8] T.A. Henzinger, P.W. Kopke, A. Puri, P. Varaiya, What's decidable about hybrid automata?, Journal of Computer and System Sciences 57 (1) (1998) 94–124.

[9] A. Puri, P. Varaiya, Decidability of hybrid systems with rectangular differential inclusions, in: G. Goos, J. Hartmanis, D.L. Dill (Eds.), Computer Aided Verification, vol. 818, Springer, Berlin, Heidelberg, 1994, pp. 95–104, https://doi.org/10.1007/3-540-58179-0_46.

[10] R. Alur, Formal verification of hybrid systems, in: 2011 Proceedings of the Ninth ACM International Conference on Embedded Software (EMSOFT), ACM Press, 2011, p. 273, https://doi.org/10.1145/2038642.2038685.

[11] R. Alur, T.A. Henzinger, P.-H. Ho, Automatic symbolic verification of embedded systems, IEEE Transactions on Software Engineering 22 (3) (1996) 181–201, doi:10/ds232r.

[12] G. Frehse, Scalable Verification of Hybrid Systems, Habilitation à diriger des recherches, Univ. Grenoble Alpes, May 2016.

[13] R. Alur, D.L. Dill, A theory of timed automata, Theoretical Computer Science 126 (1994) 183–235.

[14] J. Bengtsson, W. Yi, Timed automata: semantics, algorithms and tools, in: Advanced Course on Petri Nets, Springer, 2003, pp. 87–124.

[15] S. Tripakis, T. Dang, Modeling, verification, and testing using timed and hybrid automata, in: Model-Based Design for Embedded Systems, vol. 20091230, CRC Press, 2009, pp. 383–436, https://doi.org/10.1201/9781420067859-c13.

[16] F. Cassez, K. Larsen, The impressive power of stopwatches, in: International Conference on Concurrency Theory, Springer, 2000, pp. 138–152.

[17] B. Bérard, S. Haddad, Interrupt timed automata, in: International Conference on Foundations of Software Science and Computational Structures, Springer, 2009, pp. 197–211.

[18] Y. Osada, T. French, M. Reynolds, H. Smallbone, Hourglass automata, Electronic Proceedings in Theoretical Computer Science 161 (2014) 175–188, https://doi.org/10.4204/EPTCS.161.16.

[19] S. Akshay, B. Bollig, P. Gastin, M. Mukund, K.N. Kumar, Distributed timed automata with independently evolving clocks, in: International Conference on Concurrency Theory, Springer, 2008, pp. 82–97.

[20] K.G. Larsen, P. Pettersson, W. Yi, UPPAAL in a nutshell, International Journal on Software Tools for Technology Transfer 1 (1–2) (1997) 134–152.

[21] S. Yovine, KRONOS: a verification tool for real-time systems, International Journal on Software Tools for Technology Transfer 1 (1–2) (1997) 123–133.

[22] A. Rajhans, S. Avadhanula, A. Chutinan, P.J. Mosterman, F. Zhang, Graphical modeling of hybrid dynamics with Simulink and Stateflow, in: Proceedings of the 21st International Conference on Hybrid Systems: Computation and Control (Part of CPS Week), HSCC '18, ACM, New York, NY,

USA, 2018, pp. 247–252, https://doi.org/10.1145/3178126.3178152, http://doi.acm.org/10.1145/3178126.3178152.

[23] P. Fritzson, V. Engelson, Modelica—a unified object-oriented language for system modeling and simulation, in: European Conference on Object-Oriented Programming, Springer, 1998, pp. 67–90.

[24] C. Brooks, A. Cataldo, E.A. Lee, J. Liu, X. Liu, S. Neuendorffer, H. Zheng, HyVisual: a hybrid system visual modeler, University of California, Berkeley, Technical Memorandum UCB/ERL M 5.

[25] D. Harel, Statecharts: a visual formalism for complex systems, Science of Computer Programming 8 (3) (1987) 231–274, doi:10/b97n8k.

[26] L.P. Carloni, R. Passerone, A. Pinto, A.L. Angiovanni-Vincentelli, Languages and tools for hybrid systems design, Foundations and Trends® in Electronic Design Automation 1 (1/2) (2006) 1–193, doi:10/cxjxnq.

[27] P. Fritzson, P. Aronsson, A. Pop, H. Lundvall, K. Nystrom, L. Saldamli, D. Broman, A. Sandholm, Openmodelica – a free open-source environment for system modeling, simulation, and teaching, in: Computer Aided Control System Design, 2006 IEEE International Conference on Control Applications, 2006 IEEE International Symposium on Intelligent Control, 2006 IEEE, IEEE, 2006, pp. 1588–1595.

[28] D. Brück, H. Elmqvist, S.E. Mattsson, H. Olsson, Dymola for multi-engineering modeling and simulation, in: Proceedings of Modelica, 2002, 2002, pp. 1–6.

[29] B.P. Zeigler, Theory of Modelling and Simulation, A Wiley-Interscience Publication, John Wiley, 1976.

[30] B.P. Zeigler, J.S. Lee, Theory of quantized systems: formal basis for DEVS/HLA distributed simulation environment, in: Enabling Technology for Simulation Science II, 1998, pp. 49–59.

[31] E. Kofman, S. Junco, Quantized-state systems: a DEVS approach for continuous system simulation, Transactions of the Society for Computer Simulation International 18 (3) (2001) 123–132.

[32] H.P. Dacharry, N. Giambiasi, Formal verification with timed automata and devs models: a case study, in: Proc. of Argentine Symposium on Software Engineering, 2005, pp. 251–265.

[33] F. Bergero, E. Kofman, PowerDEVS: a tool for hybrid system modeling and real-time simulation, SIMULATION 87 (1–2) (2011) 113–132, doi:10/bwzxtb.

[34] C. Petri, Kommunikation mit Automaten, Ph.D. thesis, Rheinisch-Westfälisches Institut für Instrumentelle Mathematik 2. Universität, Bonn, 1962.

[35] L. Popova-Zeugmann, Time and Petri Nets, Springer, 2013.

[36] L. Recalde, S. Haddad, M. Silva, Continuous Petri nets: expressive power and decidability issues, in: K.S. Namjoshi, T. Yoneda, T. Higashino, Y. Okamura (Eds.), Automated Technology for Verification and Analysis, Springer, Berlin, Heidelberg, 2007, pp. 362–377.

[37] H. Alla, R. David, Continuous and hybrid Petri nets, Journal of Circuits, Systems, and Computers 8 (01) (1998) 159–188.

[38] R. David, H. Alla, On hybrid Petri nets, Discrete Event Dynamic Systems 11 (1–2) (2001) 9–40, doi:10/dp9ks6.

[39] R. David, H. Alla, Discrete, Continuous, and Hybrid Petri Nets, 2nd edition, Springer Publishing Company, Incorporated, 2010.

[40] L. Ghomri, H. Alla, Z. Sari, Structural and hierarchical translation of hybrid Petri nets in hybrid automata, in: IMACS05, 2005, p. 6.

[41] T. Bourke, M. Pouzet, Zélus: a synchronous language with ODEs, in: 16th International Conference on Hybrid Systems: Computation and Control, Philadelphia, USA, 2013, pp. 113–118.

[42] N. Halbwachs, P. Caspi, P. Raymond, D. Pilaud, The synchronous dataflow programming language LUSTRE, in: Proc. of the IEEE, 1991, pp. 1305–1320.

[43] R.B. Franca, J.P. Bodeveix, M. Filali, J.F. Rolland, D. Chemouil, D. Thomas, The AADL behaviour annex – experiments and roadmap, in: 12th IEEE International Conference on Engineering Complex Computer Systems, 2007, pp. 377–382.

[44] S. Klikovits, A domain-specific language approach to hybrid CPS modelling, Ph.D. thesis, University of Geneva, Switzerland, 2019, https://doi.org/10.13097/archive-ouverte/unige:121355, https://archive-ouverte.unige.ch/unige:121355.

[45] S. Klikovits, A. Linard, D. Buchs, CREST – a DSL for reactive cyber-physical systems, in: F. Khendek, R. Gotzhein (Eds.), 10th System Analysis and Modeling Conference (SAM2018). Languages,

Methods, and Tools for Systems Engineering, in: Lecture Notes in Computer Science, vol. 11150, Springer, 2018, pp. 29–45, https://doi.org/10.1007/978-3-030-01042-3_3.

[46] D. Broman, C. Brooks, L. Greenberg, E.A. Lee, M. Masin, S. Tripakis, M. Wetter, Determinate composition of FMUs for co-simulation, in: 2013 Proc. Int. Conf. Embed. Software, EMSOFT 2013 (Emsoft), https://doi.org/10.1109/EMSOFT.2013.6658580.

[47] M. Browne, E. Clarke, O. Grümberg, Characterizing finite Kripke structures in propositional temporal logic, Theoretical Computer Science 59 (1) (1988) 115–131, doi:10/br8284.

[48] E.M. Clarke, E.A. Emerson, Design and synthesis of synchronization skeletons using branching time temporal logic, in: Workshop on Logic of Programs, Springer, 1981, pp. 52–71.

[49] A. Pnueli, The temporal logic of programs, in: 18th Annual Symposium on Foundations of Computer Science (SFCS 1977), 1977, pp. 46–57, https://doi.org/10.1109/SFCS.1977.32.

[50] T.A. Henzinger, X. Nicollin, J. Sifakis, S. Yovine, Symbolic model checking for real-time systems, Information and Computation 111 (2) (1994) 193–244.

[51] D. Lepri, E. Ábrahám, P.C. Ölveczky, Sound and complete timed CTL model checking of timed Kripke structures and real-time rewrite theories, Science of Computer Programming 99 (2015) 128–192, doi:10/f6zwpd.

[52] D.C. Black, J. Donovan, B. Bunton, A. Keist, SystemC: From the Ground Up, Springer-Verlag New York, Inc., Secaucus, NJ, USA, 2010.

[53] A. Bariic, V. Amaral, M. Goulao, Usability evaluation of domain-specific languages, in: 2012 Eighth International Conference on the Quality of Information and Communications Technology, IEEE, 2012, pp. 342–347.

[54] A. Barisic, V. Amaral, M. Goulão, Usability driven DSL development with USE-ME, Computer Languages, Systems and Structures 51 (2018) 118–157, https://doi.org/10.1016/j.cl.2017.06.005.

[55] S. Klikovits, A. Coet, D. Buchs, ML4CREST: machine learning for CPS models, in: 2nd International Workshop on Model-Driven Engineering for the Internet-of-Things (MDE4IoT) at MODELS'18, in: CEUR Workshop Proceedings, vol. 2245, 2018, pp. 515–520, http://ceur-ws.org/Vol-2245/mde4iot_paper_4.pdf.

[56] S. Klikovits, D. Buchs, Pragmatic reuse for DSML development, Software and Systems Modeling (SoSyM) (2020), https://doi.org/10.1007/s10270-020-00831-4.

PART 3

Case studies

CHAPTER 10

Development of an IoT and WSN based CPS using MPM approach: a smart fire detection case study[☆]

Moharram Challenger[a], Raheleh Eslampanah[a], Burak Karaduman[b], Joachim Denil[a] and Hans Vangheluwe[a]
[a]University of Antwerp and Flanders Make, Antwerp, Belgium
[b]International Computer Institute, Ege University, Izmir, Turkey

Learning objectives

After reading this chapter, we expect you to be able:

- To elicit the requirements and the design of an IoT and WSN based CPS system

- To know how to use MPM for the development of the CPS

- To use FTG+PM for modelling the transformation flow and process model of the CPS system.

10.1. Introduction

This chapter discusses the design and development of a Cyber-physical System (CPS) for a fire detection system, consisting of an Internet of Things (IoT) system and Wireless Sensor Network (WSN). The proposed CPS covers the communication aspect of the system (between WSN nodes, IoT elements, and cross-platform), control aspect to interact with the physical world and system actors (using sensors and actuators), and embedded computing (by embedded software in the sensor motes, Edge, and Gateway).

There are many tools, technologies, and approaches to prevent, detect, and automatically extinguish a fire. Recently, automation mechanisms and smart building technologies are increasingly used to prevent fire spreading. Among the others, WSN [1][2][3] and IoT [4][5][6] technologies are used to propose smart solutions for CPS [7] such as fire detection system at indoor or outdoor areas.

WSN devices are developed with many features such as RPL (Routing Protocol for Low-Power devices), IPv6 support with the ability to join the sensor network instantly and operate as a repeater to extend the network by using IEEE 802.15.4 protocol, resulting to have a scalable system. IoT systems can be integrated with WSN to build a CPS

[☆] This work is result of a collaborative work between CoSys-Lab and AnSyMo Reseach Group at University of Antwerp, Ege-SERLab at Ege University, and TESLA Electronics Lab at Izmir Economy University.

Multi-Paradigm Modelling Approaches for Cyber-Physical Systems
https://doi.org/10.1016/B978-0-12-819105-7.00015-5

and benefit from its advantages. However, the CPS focuses on the control and human-in-the-loop of the system. The fire detection CPS discussed in this chapter applies this methodology for the development of the system. In this way, scalability (mesh network), flexibility (ad hoc connectivity), and low-power (using 802.15.4 protocol) characteristics of WSN are combined with the ubiquity, convenience, and cost-efficiency of the IoT system to build the required CPS.

The proposed CPS has many heterogeneous components resulting in accidental complex considering both the structure and the behaviour of the CPS. Therefore, multiple paradigm modelling approach [8] is used to address the modelling of various parts of the CPS as well as different phases of CPS development. This leads to modelling the subsystems and their behaviour explicitly at an appropriate abstraction level. Regarding the smart fire detection case study, these modelling paradigms include use case diagrams for analysis, block definition diagram for design, state-machine for modelling the behaviour, component diagram for interaction and so on.

To present the multi-paradigm models which are used along with the overall development process of the proposed system, the FTG+PM framework [9,10] is utilised and all the involved artifacts and model transformations are described. In this way, the data flow and the control flow of the system development are presented in the PM (Process Model). Furthermore, the analysis of the FTG (Formalism Transformation Graph) shows the possible improvements in the system development by finding the critical manual transformation to be (semi-)automated.

The rest of the chapter elaborates on the MPM (Multi-Paradigm Modelling) based engineering of a CPS (constitute of WSN and IoT system) for fire detection use cases. In Sections 10.2 and 10.3, the analysis and design of the system are discussed to elicit the system requirement and provide the design artifacts based on these requirements. The modelling and simulation of the designed system are presented in Section 10.4. The implementation and evaluation of the proposed CPS are elaborated in Section 10.5. Section 10.6 discusses the formalism transformation graph and detailed development process of the system. Finally, Section 10.7 summarises the chapter and Section 10.8 suggests some materials for further reading.

10.2. Requirement elicitation

The analysis and design of the fire detection system are realised using the UML and SysML principles [11]. The interaction between the system and the actors is represented by use case diagrams. Moreover, the use cases are specified with listed condition sentences. Since the system is designed for a public library, the system requirements are extracted from the general description provided by library authorities and staff.

The smart fire detection system should alert in case of temperature rising in different locations (in this case, of a library). The system considers the local temperature values (in specific spots), the building/room's general temperature, and forecast information to be able to compare the temperature data. The local temperature is the temperature collected by a WSN source node/mote locally and the building temperature is the temperature measured by a Wi-Fi module in a common area of the building/room.

For safety considerations, the system should be able to work without an Internet connection. Also, the system should be integrated with the library's legacy IT system. The system also has to inform the involved actors including "Librarian", "Security", and "Visitors" for taking different actions such as directing visitors to a safe place and call the fire department in case of a fire. Therefore, it is assumed that the system has soft real-time feedback from the actors.

To distinct different states in the system, the system alarm/state can turn to Blue, Yellow, and Red. For instance, in case of having a 5–10 degrees difference between local temperature and building/room temperature, the Yellow alarm is activated, while Red alarm is set when the local temperature is 10 degrees or above the room temperature. Figs. 10.1 and 10.2 describe Use Case diagrams of the system for Yellow and Red alarm scenarios. Since Blue alarm is a steady-state while there is no danger, the actors do not have any roles in this state.

As a representative case, the normal scenario flow for the Yellow alarm is described below:

1. System detects a heat rise and reaches the Yellow border (e.g. the difference becomes 5 degrees).
2. GUI screen becomes Yellow; this means that the local temperature is above the room temperature and it is increasing.
3. The system notifies the librarian and security to confirm the possible danger or to check for a false-positive.
4. In case of false-positive condition alarm is turned off, and the system returns to the safe condition (Blue State).
5. If one of the actors confirms the fire:
   a. The GUI screen becomes Red.
   b. The system notifies actors and visitors via a mobile app or a web form (e.g. bulletin screens or security monitors).
   c. The system notifies the fire department.
   d. The system sends the visitors and actors required information to reach a safe place to the closest exit using shortest paths.
6. If the visitors receive the fire danger notification and reach a safe place, then they click on the "I am safe" button on the app or on the web page.
   The important requirements fulfilled by the system are:
   - The system is able to detect the temperature rising and respond to that in at most 4 seconds.
   - The system is able to detect temperature rising locally.
   - The system is able to consider the local temperature, room temperature, and forecast information to decide there is a fire danger.

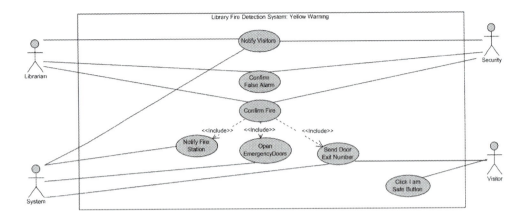

Figure 10.1 Use case diagram for Yellow alarm scenario.

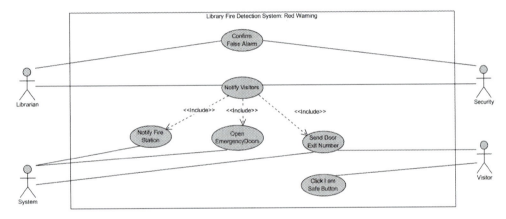

Figure 10.2 Use case diagram for Red alarm scenario.

- The system is able to state (Blue screen) when the local temperature is below the room temperature.
- The system is able to warn (Yellow screen) when the local temperature is slightly above (e.g. 5 degrees) room temperature.
- The system is able to warn (Red screen) when the local temperature is higher than room temperature (e.g. 10 degrees or more).
- The system is able to open all the emergency doors to provide fast evacuation.
- The system is able to keep logs of the temperature data, time, and date.
- The system is able to work without Internet Connection.

In Case of Fire:
- The system is able to notify actors via mobile app or web form.
- The actors are able to control the alarm.

- The system is able to show the closest emergency exit for visitors.
- The system is able to notify the fire department.

To sum up, in the system analysis phase, the description of the system is collected in the form of free text in collaboration with end-users. This document is used to elicit the system requirements. To this end, two formalisms are used: use case diagrams and use case scenarios. The use case diagram is a representation of a user/actor's interaction with the system that shows the relationship between the user and the different use cases in which the user is involved. A use case scenario specifies an important path/flow through the use case. In this study, the scenarios for normal, Yellow and Red states of the system are used, however, only the scenario for Yellow alarm is presented as an example. Finally, the transformation between the free text system description, use case diagrams and use case scenarios is done manually.

10.3. System design

This section covers the system design including the architectural design and detailed design.

10.3.1 Architectural design

In the fire detection use case, IoT and WSN paradigms are used in the physical environment to collect building/room and local temperature values and send them to the Log Manager; see Fig. 10.3. The Log Manager processes the data and sends messages to actors to confirm the fire (if the system is in the Yellow state). Upon the confirmation, the required instructions are sent to the smartphones of the visitors to direct people out of the building. The Log Manager can also send commands to the related door locks (e.g. the paths to emergency exits) to open the doors. The system also can get weather forecast information to consider in the fire detection process, which can be useful for outdoor environments. The overview of the proposed system is shown in Fig. 10.3.

Considering the WSN part, the system is designed using IRIS[1]/TmoteSky[2] motes to measure the local temperature. These motes have a low-power IEEE 802.15.4 wireless module. TmoteSky has a CC2420 antenna board with an MSP430 microprocessor, and IRIS has AVR microprocessor and uses AT86RF antenna chip.

These motes are battery-powered. They have energy-efficient microprocessors and low energy transmission protocols, leading them to be long-life battery powered components. So, this network can work without the need for external power for a long time. Also, the protocol used in this network, IEEE 802.15.4, works with mesh struc-

---

[1] https://www.memsic.com/userfiles/files/User-Manuals/iris-oem-edition-hardware-ref-manual-7430-0549-02.pdf.

[2] https://www.advanticsys.com/shop/mtmcm5000msp-p-14.html.

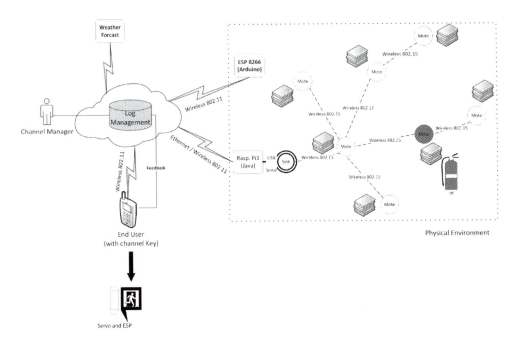

Figure 10.3 The architecture of fire detection system.

ture to deliver the messages, which sends the messages to a sink node by forwarding through neighbour nodes. As a result, it can work without an Internet connection, initial (pre-configured) topology (so-called ad hoc). This lets a WSN be scalable.

The sink node can deliver the collected local temperature values to a gateway to forward them to the Log Manager. The Log Manager can be on the cloud and their connection can be made using the Internet. The Log Manager can also be on the gateway. In the latter case, there is no need for the Internet connection. In this study the latter approach is used. Although the Internet connection is also used to collect the weather forecast information.

To realise the gateway, a RaspberryPi and a Java application running on it are used. The Java application is designed to read the serial port, to send the incoming data to the Log Manager system, to gather forecast information from a website using an API, and send notifications about fire danger to smartphones. Moreover, the Java program has a GUI which displays Yellow, Blue, or Red screens depending on the local temperature and room temperature comparison result.

On the other hand, the room/building temperature is measured by an IoT element, called ESP8266,[3] positioned at the centre of the room. It is programmed using C lan-

---

[3] https://www.espressif.com/en/products/hardware/esp8266ex/overview.

guage on Arduino IDE. If there is a fire at a place of the room, the temperature rises at around that place. To detect temperature rising, each local temperature at a specific place of a room is compared to the room temperature. To this end, the ESP8266 IoT module is used working with the IEEE 802.11 protocol based Wi-Fi transceiver.

Briefly, in the scope of this study, the WSN nodes/motes are placed on the bookshelves of a library to collect local temperature data, and the ESP8266 is positioned at the centre of the room/hall to measure the room temperature.

As the last part of the system, it includes a cloud-based information processing system called Log Manager to reduce the complexity by centralising the control mechanism and to store the temperature data. It stores forecast information, source nodes local temperature values, and ESP8266's room temperature. Moreover, the Log Manager can also make a logical comparison between room temperature and local temperatures. If the temperature rises locally, an HTTP request is sent to an ESP8266 to unlock the emergency exit door.

10.3.2 Detailed design

The detailed design of the system is realised using SysML modelling language. The design starts with a high-level view of the building blocks of the system using the Block Definition Diagram (BDD).

The BDD of the system is shown in Fig. 10.4. The local temperature values and room temperature values that are collected by Sink node and ESP8266 element are stored in a cloud system called Log Manager. Notifications about fire danger are sent to the smartphone by the Java Application using PushBullet API.[4] The emergency doors can be opened by Log Manager using HTTP requests. The Java Application and Log manager operate on the RaspberryPi that is also a gateway for the WSN part. The Log Manager has online access to stored data and can report them graphically.

As can be seen in Fig. 10.4, all elements of the fire detection system are contained by the system itself. Each element is connected with a composition relation as it has a unique role for the system. The ESP8266 has an LM35 temperature sensor and the RaspberryPi contains Log Manager, Java Application, and Notification elements. Since sink nodes are composed of many source nodes, it is also designed using a composition relation with source nodes.

The Internal Block Diagram (iBD) of the Java Application, demonstrated in Fig. 10.5, describes the data flow of the components. The main class in the Java application compares the room temperature and local temperature values collected from ESP8266 and the Sink node. Moreover, according to the temperature difference between local temperature and room temperature, the GUI screen's colour changes. At

---

[4] https://www.pushbullet.com.

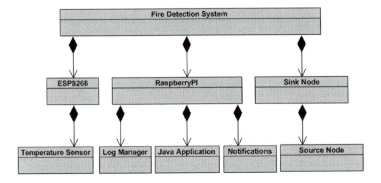

Figure 10.4 Block definition diagram of library fire detection system.

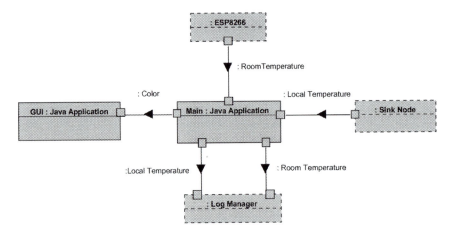

Figure 10.5 Internal block diagram of Java application.

the same time, room temperature and local temperature values are stored in the Log Manager's Database.

The state-machine diagram, Fig. 10.6, represents the behaviour of the system and describes the system's reaction to compare the results of temperature values. According to the temperature difference between the local temperature and the room temperature, the system changes its state. Since this works continuously, it always measures the temperature values and its execution can only be ended when it is shut down externally by the user. The system always returns to the Blue state (steady-state).

While the local temperature value is below the room temperature the system stays in the Blue state, i.e., the system is in a safe state. The system switches to the Yellow alarm state when local temperature approximates to the room temperature. When the system is switched to this state the Yellow alarm scenario is realised (see Fig. 10.1). If the fire warning alarm is controlled and it is a false alarm, it returns to the safe state.

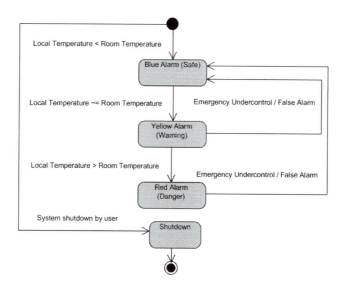

Figure 10.6 State machine diagram of library fire detection system.

If the local temperature continues rising and goes above the room temperature (e.g. 5 degrees higher), the system switches to the Red alarm state. When the system is switched to this state the Red alarm scenario is realised (see Fig. 10.2). If the fire danger is under-controlled or it is a false alarm, it returns to the safe state.

To provide a low-level view of the interaction of components, a component diagram is designed for the fire detection system (see Fig. 10.7). In this way, communication technology and the interfaces that connect the system components to each other are specified explicitly. This component diagram shows the system elements' physical connections and the technologies that combine the system components. The Blue ball shows an interface provided as an output and the semi-circle shows an input interface required for the system. In this case, an input is provided by the system to sample the room temperature. The room temperature can be receipted by the surface of the LM35 temperature sensor. Next, the internal structure of the temperature sensor is an interface to transform a physical phenomenon to the interpret-able digital data. Therefore, the temperature data is receipted using sensor components and this data is outputted by a specific pin of the temperature sensor.

Using a wire, the output pin of the temperature sensor is connected to the analog pin of the ESP8266. In this way, the temperature data is delivered to the ESP8266. Using the sockets and physical antenna, compliance to IEEE 802.11, the room temperature is transmitted to the RaspberryPi.

Since RaspberryPi is also used as a gateway for the WSN part of the system, it receives the local temperature via the serial port. The Log Manager element is associated

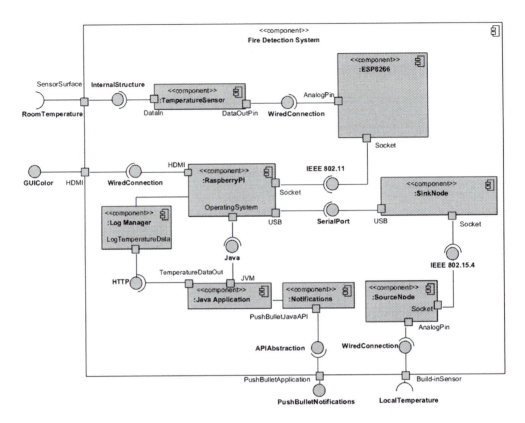

Figure 10.7 Component diagram of library fire detection system.

with the RaspberryPi and it both receives the data and sends the HTTP requests using RaspberryPi capabilities.

A Java application runs on the RaspberryPi, to handle the tasks in RaspberryPi side. For example, it handles the notifications by sending them to smartphones of the library visitors via the PushBullet application.

GUI's colour is changed by the comparison result of the local temperature and room temperature and the colour data is conducted by the HDMI connection.

In conclusion, during the design phase, the following paradigms are used: block definition diagram, internal block diagram, state-machine diagram, and component diagram. To this end, the paradigms and artifacts in the system analysis phase are used. For example, the use scenarios are used to provide the state-machine diagrams. It is worth to note that the transformations are fulfilled manually. From these design models, the elements can be selected for the system, the wired and wireless network setup can be designed, and these setups and elements can be verified in the design phase.

10.4. Modelling and simulation

Different aspects of the system are modelled in domain-specific modelling and simulation tools using related paradigms to define components, connections, and validate the design choices. The starting point to model the system is its sensors and actuators that provide an input and output for the fire detection system. To provide a fully-functional system, the physical interaction with environment as well as computations must be provided.

To this end, the mathematical power consumption model of the servo motor is given. A servo motor such as SG90[5] needs the electrical power to pull weight at a certain distance. In our case, it has to pull the door lock. Based on the SG90 specs, this servo motor has power consumption for instantly lifting a weight and power consumption while idling as follows:

Servo motor power consumption model for SG90:
* Lifting current consumption for 2.5 kg load is 270 mA, where the input voltage is 5 volts.
* Idling current consumption for 2.5 kg load is 6 mA, where the input voltage is 5 volts.

This information can be used in the calculation of the system's power consumption and selecting an efficient servo motor for the system. Because the servo motor has to provide enough torque to pull the door lock, the simulation of the servo motor has to be made to observe the duty-cycle and rotational movement of the servo motor.

Servo motor duty-cycle and rotational movement for SG90:
* Position "0" (1.5 ms pulse) centres the servo motor.
* Position "90" (~2 ms pulse) turns the servo all the way to the right.
* Position "-90" (~1 ms pulse) turns the servo all the way to the left.

The rotation behaviour of the servo motor is simulated by sending pulses according to given intervals above. A pulse generation element is used to create Pulse-Width-Modulation signals. According to the above specifications, the servo motor is simulated in Proteus simulator[6] (see Fig. 10.8).

In this case study, the servo motor is fed by the limitless power source, so, the power consumption is not critical. Even if electricity goes out in the building door locks can be released automatically using battery power. However, the servo motor still has to produce enough torque (T) to pull the door lock. Therefore it is important to consider specs of the selected servo motor. SG90's requirements are as follows to produce enough

---

[5] http://www.ee.ic.ac.uk/pcheung/teaching/DE1_EE/stores/sg90_datasheet.pdf.
[6] https://www.labcenter.com.

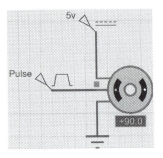

Figure 10.8 Servo motor's simulation.

torque (T).

$$T_{\text{nominal}} = 2.5_{\text{kg/cm}}$$

where

$$V_{\text{supply}} = 5_{\text{volts}} \text{ and } I_{\text{supply}} = 270_{\text{mA}}$$

The above equation shows that SG90 must be fed by a power source that can give at least 1.35 watt power. Therefore, during the implementation of the fire detection system, 1.5 watt power source is used. Since the power requirements are provided, enough torque can be produced by the SG90 servo motor. The nominal torque value shows that the servo motor can pull a 2.5 kg weight where there is a 1 cm distance between servo and the weight. In this case study, a 500 g door lock is positioned at 1 cm away from the servo motor and it is pulled successfully.

In this use case, after modelling the required components, the simulation of WSN is realised using the Cooja simulation framework[7] for both single-hop topology and multi-hop topology (see Figs. 10.9 and 10.10, respectively). This helps us to analyse the properties such as message routing and bottlenecks in the network. This analysis is very important in ad hoc networks, as it is possible that a node does not deliver a message, e.g. when the neighbour node is far away from the node's transmission range. The single-hop and multi-hop networks are created and motes for the sink node and source nodes are tested in Cooja.

Fig. 10.9 demonstrates an established network between two sensor nodes. In a single-hop network, the nodes can send and receive data to the server within a 120-meter range if there is no obstacle to reduce the transmission distance. However, in indoor areas, there are many obstacles that reduce the transmission range to 50-meter. In this sense, the multi-hop topology is more preferable to increase the coverage.

The source nodes in a multi-hop network can transfer data packets using a neighbour node that is in the coverage distance. Therefore, the data packets are sent to the

---

[7] https://github.com/contiki-os/contiki/wiki/An-Introduction-to-Cooja.

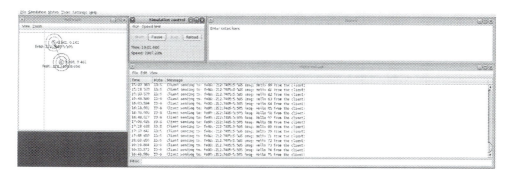

Figure 10.9 Source node sending data to a Sink node in a single-hop WSN.

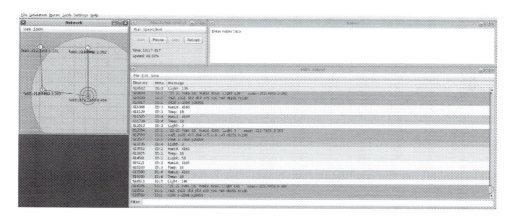

Figure 10.10 Successful data transmission in a multi-hop WSN.

sink node by hopping node-by-node. Fig. 10.10 represents a topology where node 4 transmits its data packet to the sink node (node 1) via selecting one of its neighbour nodes, node 2 or node 3.

If a node is not in the range of the source node, and if it has neighbour nodes then it transfers the data packet to its neighbour/repeater node and this cycle continues until the data packet is received by server node.

As a result of simulating the single-hop and multi-hop networks using Cooja, it is decided to use a multi-hop network due to wide coverage capability. This is especially useful in indoor environments. Moreover, the prototype codes for sensor sampling and message routing are simulated before implementing the system.

To summarise, for modelling and simulation of the fire detection system, the servo motors (as one of the actuators of the system) are modelled using power consumption equations. Also, the networking aspect of the system is modelled and simulated using Cooja.

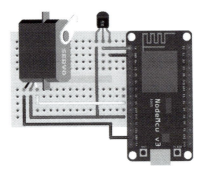

Figure 10.11 ESP8266 and LM-35 temperature sensor wiring.

10.5. Implementation

The system is implemented based on the requirements, models, and simulation results that are mentioned in the previous sections. The implementation is realised for each sub-parts of the system including hardware, network, software, and log manager.

10.5.1 Hardware setup

The wiring schematic of the ESP8266 Wi-Fi transceiver and the LM35 temperature sensor is demonstrated in Fig. 10.11. In this case study, LM35's input voltage is 3.3 V with the current drain of 56 μA. The LM35 senses the heat via its surface. Then the heat is converted to internal resistance. For each rising of one degree Celsius, the LM35's output voltage rises with 10 mV. The output pin of the LM35 is connected to the ADC input pin of the ESP8266. In this way, the room temperature is sampled by the ESP8266.

Moreover, another ESP8266 controls the servo motor using the Pulse-Width-Modulation technique. Fig. 10.11 shows how the servo motor and LM35 are connected to a single ESP8266. However, in the implementation, the servo and LM35 are connected to different ESP8266's, as the ESP8266 with LM35 temperature is positioned at the centre of the room and the ESP8266 with servo motor is positioned on the door to open the lock.

In addition, the servo motor is fed by 5V to produce the required torque to pull the door lock. The ESP8266 with the LM35 sensor is also used with the deep-sleep mode in case there is no unlimited power source at the centre of the room. The device is woken up each minute periodically, it samples the room temperature and sends it to the Log Manager. The ESP8266's program is implemented with C language using the Arduino IDE.

Considering the WSN, the nodes do not need a specific hardware setup, as they work in an ad hoc way and they have built-in temperature, humidity, and light sensors.

They only need to be programmed using an operating system such as ContikiOS which is discussed in the next section.

10.5.2 Software development

To develop a multi-hop network, the ContikiOS RPL library is used [12]. Two motes (the code for sensor node) are developed for the sink node and source nodes. The RPL library is used in the sink node to create a network tree and advertise itself as the root node. The source nodes receive these messages and broadcast them to the neighbour nodes. When the RPL network tree is created, the data packets can be sent to the sink node by hopping node-by-node. The sink node must re-pair the node tree in case of changing the location of a source node, adding a new node, and removal of a node from the network. Source node software is implemented to sample the temperature values and transmit them to the sink node. Listing 10.1 shows an excerpt from the RPL code of the sink node.

Listing 10.1: An excerpt from the RPL code.

```
1   if ( root_if != NULL) {
2       rpl_dag_t *dag = rpl_set_root  (RPL_DEFAULT_INSTANCE,
3           ( uip_ip6addr_t *) ipaddr ) ;
4       uip_ip6addr ( &ipaddr, 0xaaaa, 0, 0, 0, 0, 0, 0, 0);
5       rpl_set_prefix  (dag,&ipaddr, 64);
6       printf ("created a new RPL network");
7   } else   printf ("failed to create a new RPL Network");
8   if (ev == tcpip _event)  tcpip_handler () ;
9   else   rpl_repair_root (RPL _DEFAULT _INSTANCE);
```

Rising the temperature causes a Red or Yellow alarm state and a notification message is sent to the users. To achieve this, PushBullet[8] Java API is used. In this case study, PushBullet is used to send the notifications in case of emergency and receive the feedback from the uses (e.g. "I am Safe" message using a specific button). Listing 10.2 shows an excerpt from the API code for PushBullet.

Listing 10.2: An excerpt from the PushBullet API.

```
1   public static void sendNotification ( String token){
2       throws PushbulletException {
3       PushbulletClient   client = new PushbulletClient (api_key) ;
4       String result = client .sendNote (null, "Fire Detection",
5           "There is a FIRE!");
6   }
```

---

[8] https://github.com/salahsheikh/jpushbullet.

In the ESP8266 side, a web-server is created for the ESP8266 with servo motor. The web-server is programmed to listen to incoming requests on port 80. When a request is made to the URI address of "/open_door", it turns the servo motor to pull the door lock. The ESP8266 with LM35 sensor is operated in deep-sleep mode. Each minute, it wakes up, samples the room temperature, and sends it to the Log Manager. Listing 10.3 shows part of the software code for the ESP8266 with LM35 sensor.

Listing 10.3: An excerpt from the ESP8266 with LM35.

```
1   void setup () { connectWifi () ; }
2   void loop () {
3       ESP.deepSleep(sleepTime*1000000, WAKE_RF_DEFAULT);
4       sendTeprature () ;
5       ...
6   }
7   void sendTeperature () {
8       if ( client .connect( server , 98)) {
9           client . print (F("POST "));
10          client . print ("temp= "+(String) temp1_i+" ") ;
11      }
12  }
```

10.5.3 Log manager

The local temperatures are sent to the sink node. Then, the sink node sends this data to the gateway, Java application. Finally, the gateway sends it to the Log Manager for storage, processing, and event handling. On the other hand, the room temperature, collected using ESP8266, is directly sent to the Log Manager.

If any local temperature value becomes higher than the room temperature value, then the Log Manager sends an HTTP request to the ESP8266 with a servo motor to open the emergency door. It also sends the required notifications. Additionally, the Log Manager can produce the log reports with the date and time graphically. Fig. 10.12 shows a new channel created for the fire detection system to log the room temperature and local temperature values.

The Log Manager is designed to respond to the acquired data and its related events. This is realised by the event definition mechanism in the Log Manager. For each event, some actions can be taken, such as sending an SMS or emails to a group of people, and/or making an HTTP request to another device. Fig. 10.13 illustrates the callback definition to make an HTTP request and event creation.

The logged temperature data can be viewed graphically to analyse and find anomalies; see Fig. 10.14. To generate this graph, a query is required in the Log Manager.

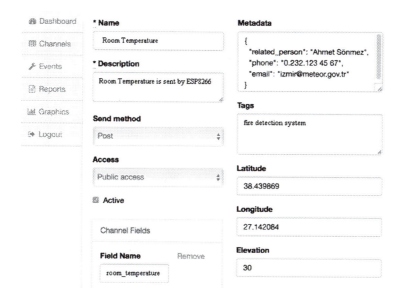

Figure 10.12 Creating a new channel in the Log Manager for fire detection system.

Figure 10.13 Event definition and callback definition.

10.5.4 Testing and verification

The fire detection system is tested as follows. A node temperature sensor has been heated by a lighter and the response time is observed. When the temperature goes up to the room temperature, the Log Manager system sends a notification to the defined user. Moreover, the Log Manager sends an HTTP request to ESP8266 with a servo motor. The servo motor's actuation is observed and it successfully pulled the door lock. At the same time, incoming request to ESP8266's is printed to the screen. The outputs of the ESP8266 are read by serial communication. PushBullet's notification message is viewed by the smartphone (see Fig. 10.15).

Furthermore, the logs of the Log Manager system are viewed to verify each node that has sent a local temperature value. Moreover, ESP8266 with the LM35 sensor value is also printed to the screen and observed by the serial monitor. The temperature value is compared with a thermometer and they were the same.

Figure 10.14 Graphical representation of the stored data.

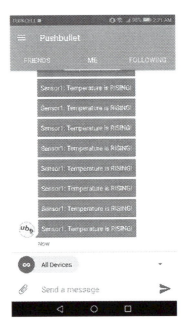

Figure 10.15 Warning messages about temperature rising.

10.6. Multi-paradigm development process in FTG+PM framework

The fire detection system is a CPS interacting with both humans and sub–systems in case of fire danger. The system is implemented according to the provided requirements.

Moreover, the system is represented using SysML diagrams and some other modelling languages. In the development of the fire detection system, various paradigms are used. So, the Multi-Paradigm Modelling approach is considered in the implementation of this system and the Formalism Transformation Graph and Process Model (FTG+PM) is utilised to describe all the involved formalisms and artifacts as well as model transformations [13]. In this way, system details are shown with data flow and the control flow in the PM. Further, the model of the FTG shows us the transformations of the formalism. The FTG and PM for the fire detection system are demonstrated in Figs. 10.16 and 10.17, respectively and they are elaborated in this section.

In general, the level of abstraction is decreased from top to bottom in PM. At the top level, the development process starts by requirement gathering and it ends when the integration test is succeeded. After the design step, the number of activities increases as the abstraction level decreases. The steps are declared as transformation definitions in the FTG and they are instantiated as activities in the PM. FTG model describes modelling languages and the transformations which are available to engineering in a selected domain. Thus, the domain expert can follow the transformation steps and data flow. The instantiated activities are represented. Artifacts are input and/or output of the activities in PM. For example, any requirement which had written in free text is the source for modelling this requirement. Moreover, according to the activity of Modelling requirements, two artifacts are derived as Use Case Specification and Use Case Diagram.

10.6.1 The Formalism Transformation Graph (FTG)

FTG of the fire detection system starts from requirements that are written in textual natural language. These requirement descriptions are the artifacts for the Modelling Requirements activity. At the end of this activity, system requirements are modelled in two artifacts shown in Fig. 10.16. Use Case Specifications and Use Case Diagrams can be refined until domain experts agree on the case. Thus, they can begin Architectural Design. The design can be done according to Use Case Specifications and Diagrams to meet the requirements.

The Architectural Design is represented by SysML diagrams. In this way, the system's architecture and sub-process interaction are described in detail. According to these diagrams, System's Network is Designed and CPS Elements are selected. Moreover, Wired and Wireless Connections are considered separately during Network Design due to the fact that these wired and wireless elements are used in different activities. At the end of these steps, all designs are combined and transformed into System Setup.

The system components that are decided to be used in the implementation can be verified using System Setup. This process is going to be elaborated in detail in PM. When all Design steps are finished the System Setup becomes clear. According to this setup, System Components are modelled. These components and their connections are the input for system Schematics. Before the implementation of these Schematics, it can

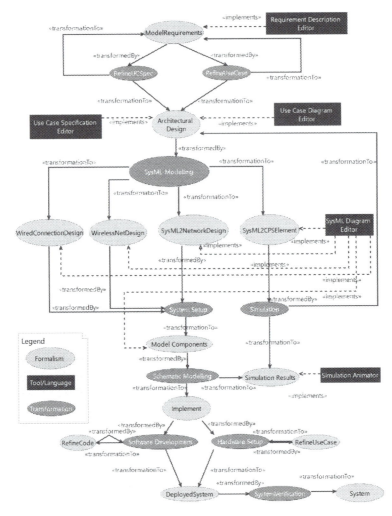

Figure 10.16 The formalism transformation graph of smart fire detection system in FTG+PM framework.

be Simulated. Selected and verified elements are used in Simulation. To have more consistent results, the elements that are supported by the Simulator can be placed in the Simulation model. As satisfactory results are produced in Simulation, Schematics of the System becomes valid and clear. Therefore, they can be implemented.

In the implementation, the software prototype (code) and the hardware prototype are produced to be set up. They can be refined until executable and persistent results are achieved. This refinement covers manually control of hardware setup and uses case match. In code refinements process, when implemented code is valid after several manual checks and compilations, code can be tested for deployment.

If hardware setup and software code satisfy use case conditions, the overall artifacts turn out to a Fire Detection System. However, this system can have bugs and can contain any improper connection. Therefore, it can be verified by testing. To verify the system, test cases can be generated and run on the system.

10.6.2 The Process Model (PM)

In the process model of the fire detection system (see Fig. 10.17), the control flow represents the sequence of activities and the data flow represents input(s) or output(s) of data for/from activity. Also the languages in the FTG Model are instantiated in PM as Objects.

The System Description is given in Free Text. This can be any native language, e.g. English in this case study. This data is input to model requirements and when this activity ends, four artifacts are instantiated as use cases. In fire detection system, Blue, Yellow and Red alarm represents three states and for each state, use cases are modelled as presented in Section 10.2. Then, according to the domain expert or use case, one or more specifications can be instantiated.

Control flow goes next to Architectural Design (Fig. 10.3), and a sketch of the system can be drawn to give a high-level view of the system. Then the system is designed in detail using SysML diagrams. Block Definition Diagram and Component Diagram are instantiated and they are represented using SysML in Figs. 10.4 and 10.7.

Control flow is separated into two activities. According to the Block Definition Diagram, a system network can be designed and Internal Block Diagram can be derived as data. In iBD, a domain expert can decide which component sends data using wired or wireless medium. At the end of the design, "Wired Connection" and "Wireless Network Design" data are gained.

According to the Component Diagram, system elements can be chosen. For example, if a temperature value needs to be acquired, a temperature sensor needs to be used in the system. However, there are various temperature sensors in the market which operate in different temperature ranges and with different precision. There might be one or more candidate sensor(s) and it is required to be verified that a correct temperature value can be measured. In AND decision, if elements are verified and designs are complete, they can be modelled. Modelling wired components is the action where wire, pin connections, and pin numbers are decided. For example, the domain expert decides whether two analog pins or three pins of I2C sensors are used (see Fig. 10.11). Modelling of Wireless Components can be taught as deciding on the number of wireless nodes and placing them in correct positions, deciding communication ports, address and antenna power, etc. After these activities, the system can be modelled and simulated.

In this study, the model of wired components is simulated in Proteus and the model of wireless nodes is simulated in Cooja as mentioned in Section 10.4. When both of the simulation results are satisfactory, each component of the system is implemented.

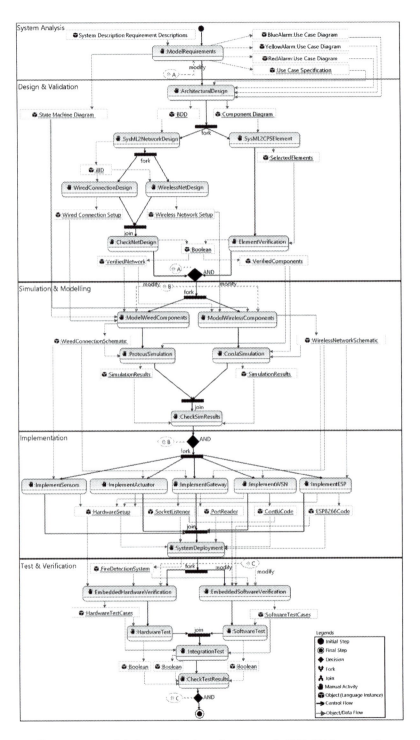

Figure 10.17 The process model of smart fire detection system in FTG+PM framework.

In the implementation of the WSN and ESP8266 C-Codes are developed and in the implementation of the gateway Java code is developed. Implementation of sensors and actuators is in the form of physical cable connections (e.g. model of wired components). These are named Hardware Setup which is corresponded to components such as ESP8266 with servo motor, ESP8266 with LM35, and TmoteSky motes with built-in temperature sensors.

After the implementation phase, the implemented software codes are flashed into the hardware components, so the system can be deployed. However, this is an untested system and may have bugs, wrong connections or defective components.

The system can be tested by dividing into two parts, respectively, Embedded Software and Embedded Hardware. Both parts need to be tested and verified. In activities of verification, test cases can be produced. If these cases are passed successfully then the system is accepted and the process is finalised.

10.7. Summary

In this chapter, the design and development of an IoT and WSN based fire detection system as a Cyber-physical system are discussed. The development of such a system is complex due to the various technologies involved and due to the variety of components which can be used. To tackle this complexity, a Multi-Paradigm Modelling (MPM) approach is used in the design and development of this system.

In the scope of the fire detection case study, the analysis and design of the fire detection system are realised using the SysML. The system is designed for a public library. The requirements of the system are extracted from the general description provided by library authorities and staff.

To present the system using the MPM approach, the Formalism Transformation Graph and Process Model (FTG+PM) are utilised. The PM model helps to provide the data flow and the control flow of the system development and the analysis of the FTG shows the possible improvements for the system by finding the critical manual transformation to be (semi-)automated.

10.8. Literature and further reading

In the study of [14], WSN devices are used along with GPS technology to find the location of a fire. The collected data is sent to the base station directly and this real-time data can be accessed by web browsers. In [15], the authors use the ad hoc and multi-hop network (using TinyOS [15]) aimed to detect wildfires for South Korea mountains and send alarm messages using the shortest path algorithm. In [16], the WSN devices with their smoke, humidity, and temperature sensors are used to communicate on Mesh

Network using MICA-2 Sensor Motes to detect fire. In [17], photo-detectors and image processing techniques are used to detect fire.

To benefit from the scalability, flexibility, and low-power characteristics of WSNs and the ubiquity, convenience and cost-efficiency of IoT systems, [18] and [19] have combined WSN and IoT technologies [6,20,21] to build a CPS for fire detection.

In [22], smart fire detection has been developed using a model-driven approach. To this end, a Contiki metamodel in [23] is extended to include elements of Wi-Fi connectivity modules (such as ESP8266), IoT Log Manager, and information processing components (such as RaspberryPi). Based on this new metamodel, a domain-specific modelling environment [24] is developed in which visual symbols are used and static semantics (representing system constraints) are defined. Similarly model-driven engineering approach [25] is applied on TinyOS [26] and RIOT [27] for the development of WSNs, as well their analysis [28].

Finally, an interesting idea to extend these smart systems, from the intelligence point of view, is to utilise intelligent agents [29] and multi-agent systems [30] in the development of IoT systems.

Acknowledgements

The authors would like to thank the COST Action networking mechanisms and support of IC1404 Multi-Paradigm Modelling for Cyber-Physical Systems (MPM4CPS). The COST is supported by the EU Framework Programme Horizon 2020.

References

[1] K. Bouabdellah, H. Noureddine, S. Larbi, Using wireless sensor networks for reliable forest fires detection, Procedia Computer Science 19 (2013) 794–801.

[2] S.H. Javadi, A. Mohammadi, Fire detection by fusing correlated measurements, Journal of Ambient Intelligence and Humanized Computing 10 (4) (2019) 1443–1451.

[3] S. Arslan, M. Challenger, O. Dagdeviren, Wireless sensor network based fire detection system for libraries, in: Computer Science and Engineering (UBMK), 2017 International Conference on, IEEE, 2017, pp. 271–276.

[4] M. Wang, G. Zhang, C. Zhang, J. Zhang, C. Li, An IoT-based appliance control system for smart homes, in: 2013 Fourth International Conference on Intelligent Control and Information Processing (ICICIP), 2013, pp. 744–747.

[5] D. Kang, M. Park, H. Kim, D. Kim, S. Kim, H. Son, S. Lee, Room temperature control and fire alarm/suppression IoT service using MQTT on AWS, in: 2017 International Conference on Platform Technology and Service (PlatCon), 2017, pp. 1–5.

[6] L. Ozgur, V.K. Akram, M. Challenger, O. Dagdeviren, An IoT based smart thermostat, in: 2018 5th International Conference on Electrical and Electronic Engineering (ICEEE), IEEE, 2018, pp. 252–256.

[7] R.F. Wills, A. Marshall, Development of a Cyber Physical System for Fire Safety, Springer, 2016.

[8] M. Amrani, D. Blouin, R. Heinrich, A. Rensink, H. Vangheluwe, A. Wortmann, Towards a formal specification of multi-paradigm modelling, in: 2019 ACM/IEEE 22nd International Conference on Model Driven Engineering Languages and Systems Companion (MODELS-C), IEEE, 2019, pp. 419–424.

[9] L. Lúcio, S. Mustafiz, J. Denil, H. Vangheluwe, M. Jukss, FTG+PM: an integrated framework for investigating model transformation chains, in: F. Khendek, M. Toeroe, A. Gherbi, R. Reed (Eds.), SDL 2013: Model-Driven Dependability Engineering, Springer, Berlin, Heidelberg, 2013, pp. 182–202.

[10] M. Challenger, K. Vanherpen, J. Denil, H. Vangheluwe, FTG+PM: describing engineering processes in multi-paradigm modelling, in: Paulo Carreira, Vasco Amaral, Hans Vangheluwe (Eds.), Foundations of Multi-Paradigm Modelling for Cyber-Physical Systems, Springer, 2020, pp. 259–271, Ch. 9.

[11] S. Friedenthal, A. Moore, R. Steiner, A Practical Guide to SysML: the Systems Modeling Language, Morgan Kaufmann, 2014.

[12] B. Pavković, F. Theoleyre, A. Duda, Multipath opportunistic RPL routing over IEEE 802.15.4, in: Proceedings of the 14th ACM International Conference on Modeling, Analysis and Simulation of Wireless and Mobile Systems, 2011, pp. 179–186.

[13] I. Dávid, J. Denil, K. Gadeyne, H. Vangheluwe, Engineering process transformation to manage (in) consistency, in: 1st International Workshop on Collaborative Modelling in MDE (COMMitMDE 2016), 2016, pp. 7–16.

[14] D.M. Doolin, N. Sitar, Wireless sensors for wildfire monitoring, in: Smart Structures and Materials 2005: Sensors and Smart Structures Technologies for Civil, Mechanical, and Aerospace Systems, vol. 5765, International Society for Optics and Photonics, 2005, pp. 477–485.

[15] B. Son, Y.-s. Her, J.-G. Kim, A design and implementation of forest-fires surveillance system based on wireless sensor networks for South Korea mountains, International Journal of Computer Science and Network Security (IJCSNS) 6 (9) (2006) 124–130.

[16] Z. Chaczko, F. Ahmad, Wireless sensor network based system for fire endangered areas, in: Information Technology and Applications, 2005. ICITA 2005. Third International Conference on, vol. 2, IEEE, 2005, pp. 203–207.

[17] P. Cheong, K.-F. Chang, Y.-H. Lai, S.-K. Ho, I.-K. Sou, K.-W. Tam, A ZigBee-based wireless sensor network node for ultraviolet detection of flame, IEEE Transactions on Industrial Electronics 58 (11) (2011) 5271–5277.

[18] B. Karaduman, M. Challenger, R. Eslampanah, ContikiOS based library fire detection system, in: 2018 5th International Conference on Electrical and Electronic Engineering (ICEEE), 2018, pp. 247–251.

[19] B. Karaduman, T. Asici, M. Challenger, R. Eslampanah, A cloud and Contiki based fire detection system using multi-hop wireless sensor networks, in: Proceedings of the Fourth International Conference on Engineering & MIS 2018, ICEMIS '18, ACM, New York, NY, USA, 2018, pp. 66:1–66:5.

[20] E. Türk, M. Challenger, An android-based IoT system for vehicle monitoring and diagnostic, in: 2018 26th Signal Processing and Communications Applications Conference (SIU), IEEE, 2018, pp. 1–4.

[21] N. Karimpour, B. Karaduman, A. Ural, M. Challengerl, O. Dagdeviren, IoT based hand hygiene compliance monitoring, in: 2019 International Symposium on Networks, Computers and Communications (ISNCC), IEEE, 2019, pp. 1–6.

[22] T.Z. Asici, B. Karaduman, R. Eslampanah, M. Challenger, J. Denil, H. Vangheluwe, Applying model driven engineering techniques to the development of Contiki-based IoT systems, in: 1st International Workshop on Software Engineering Research & Practices for the Internet of Things, Co-located with ICSE 2019, SERP4IoT'19, Montreal, QC, Canada, IEEE, 2019, pp. 25–32.

[23] C. Durmaz, M. Challenger, O. Dagdeviren, G. Kardas, Modelling Contiki-based IoT systems, in: OASIcs–OpenAccess Series in Informatics, vol. 56, Schloss Dagstuhl-Leibniz-Zentrum fuer Informatik, 2017, pp. 5:1–5:13.

[24] G. Kardas, Z. Demirezen, M. Challenger, Towards a DSML for semantic web enabled multi-agent systems, in: Proceedings of the International Workshop on Formalization of Modeling Languages, 2010, pp. 1–5.

[25] E.A. Marand, E.A. Marand, M. Challenger, DSML4CP: a domain-specific modeling language for concurrent programming, Computer Languages, Systems and Structures 44 (2015) 319–341.

[26] H.M. Marah, R. Eslampanah, M. Challenger, DSML4TinyOS: code generation for wireless devices, in: 2nd International Workshop on Model-Driven Engineering for the Internet-of-Things (MDE4IoT), 2018, pp. 509–514.

[27] B. Karaduman, M. Challenger, R. Eslampanah, J. Denil, H. Vangheluwe, Platform-specific modeling for RIOT based IoT systems, in: IEEE/ACM 42nd International Conference on Software Engineering Workshops (ICSEW'20), Virtual Event, Republic of Korea, June 2020, pp. 639–646.

[28] B. Karaduman, M. Challenger, R. Eslampanah, J. Denil, H. Vangheluwe, Analyzing WSN-based IoT systems using MDE techniques and Petri-net models, in: 4th International Workshop on Model-Driven Engineering for the Internet-of-Things (MDE4IoT), Co-Located With Software Technologies: Applications and Foundations (STAF 2020), Virtual Event, Norway, 22-26 June 2020, pp. 35–46.

[29] B.T. Tezel, M. Challenger, G. Kardas, A metamodel for Jason BDI agents, in: 5th Symposium on Languages, Applications and Technologies (SLATE'16), Schloss Dagstuhl-Leibniz-Zentrum fuer Informatik, 2016, pp. 1–9.

[30] V. Mascardi, D. Weyns, A. Ricci, C.B. Earle, A. Casals, M. Challenger, A. Chopra, A. Ciortea, L.A. Dennis, Á.F. Díaz, et al., Engineering multi-agent systems: state of affairs and the road ahead, ACM SIGSOFT Software Engineering Notes 44 (1) (2019) 18–28.

CHAPTER 11

Development of industry oriented cross-domain study programs in cyber-physical systems for Belarusian and Ukrainian universities

Anatolijs Zabasta[a], **Nadezda Kunicina**[a], **Oksana Nikiforova**[b], **Joan Peuteman**[c], **Alexander K. Fedotov**[d], **Alexander S. Fedotov**[d] and **Andrii Hnatov**[e]

[a]Institute of Electrical Engineering and Electronics, EEF Riga Technical University, Riga, Latvia
[b]Riga Technical University, Riga, Latvia
[c]KU Leuven, Brugge, Belgium
[d]Energy Physics Department, Physics Faculty, Belarusian State University, Minsk, Belarus
[e]Kharkiv National Automobile and Highway University, Kharkiv, Ukraine

11.1. Introduction

The impact of the Fourth Industrial Revolution (Industry 4.0) [32] is not restricted to industrial production processes. The rise of Industry 4.0 related technology also has a large number of applications outside the traditional industry and includes the everyday life of an increasing number of people. The rise of Industry 4.0 related technology is driven by the availability of the internet, communication technology and (real time) computing power of processors. In the past, prototypes of Cyber-Physical Systems (CPS systems) and products were mainly physical objects but today they include a software level using the increasing computing power of a large range of processors.

In order to speed up the development of new CPS products and reduce their design costs, modelling and simulating the controlled processes gain importance. These modelling and simulation approaches also help to improve the product quality. More precisely, the development of new products in a virtual environment based on real data, but at a much lower cost and with less risk, becomes possible. Therefore, modelling becomes an integral part of all innovation processes [1].

The technical evolutions in the CPS world are ever accelerating but, in general, adapting the curricula in a university is a rather slow process. The evolutions in the industry and the research in the academic world do not always correlate well [2]. This implies that bridging the gap between industry needs and educational output, in terms of the prospective researchers and CPS engineers, is always a challenge. In this context, one of the main goals is the introduction of new up-to-date curricula to avoid outdated curricula and teaching contents. It is also important that students gain research

Multi-Paradigm Modelling Approaches for Cyber-Physical Systems
https://doi.org/10.1016/B978-0-12-819105-7.00016-7

and design oriented skills [3] which allow them to apply the academic and technical knowledge when designing and introducing CPS systems.

Additionally, students seeking for a CPS educational program have diverse educational backgrounds with different expertise. Due to this diversity, it is difficult to teach to these students the challenging CPS design related topics. It is also important to identify what tools and methods can empower CPS developers to overcome the technical problems involved in creating new systems [4]. Engineering of CPS requires skills, which include a very broad range of technologies, i.e. a multidisciplinary approach is needed and cross–domain study programs must be developed. To develop the SE (System Engineering) of a CPS program, it is necessary to build a series of courses, which satisfy these needs. Effectively integrating the skills of hardware, software and systems engineering is mandatory.

The Riga Technical University (RTU), in cooperation with the other members of Multi-Paradigm Modelling for CPS MPM4CPS COST (European Cooperation in Science and Technology) Action [5], uses their knowledge, experience and methods acquired in the Action in order to verify the viability of the approach and methods developed in the Action for creating industry-focused curricula in Higher Educational Institutions (HEIs) in Partner Countries (PCs, i.e. non–EU countries).

In this research, we discuss how the COST team efforts and methods towards analysis of tendencies, industry needs, and use of modern educational practices have been made by the ERASMUS+ team [6], [7], [8], [29], [30], [31] in order to create an industry-oriented curriculum on CPS modelling in HEIs of Belarus and Ukraine. We discuss the results of a survey of Belarusian and Ukrainian stakeholders on the prospected CPS related topics. We discuss a unique experience, i.e. the implementation of ideas and perceptions of COST participants on Belarusian and Ukrainian HEIs. As it was the case in the previous projects, exchanging experiences among all partners is very important [30].

First of all, we provide a short review of labour markets of Belarus and Ukraine, which is based on long-term cooperation between partners' universities and employers (see Chapter 3). Then we recall the survey among the partners of the COST Action in order to identify and classify the competencies, which partners recognise as indispensable and most relevant related to particular areas of CPS [9] (see Chapter 4).

Taking into account techniques and tools validated in COST Action, we gathered a consortium of high-tech Belarusian industry and research institutions as well as Ukrainian transport and mining industrial enterprises. We recognise these institutions as the stakeholders of the ERASMUS+ Physics (2015–2018) and CybPhys (2019–2022) projects, in order to identify a CPS expert's profile suitable for the Belarusian and Ukrainian labour market needs (see Chapter 5).

Based on outcomes of both surveys, a working group consisting of academicians and researchers of European, Belarusian and Ukrainian universities, made a number of recommendations and a list of courses on the modelling of CPS. The courses intend to educate bachelor and master students. We are going to validate this approach and methods at the new ERASMUS+ project "Development of practically-oriented student-centred education in the field of modelling of Cyber-Physical Systems", "CybPhys" in order to see how newly developed curricula allow one to meet the needs of the Belarusian and Ukrainian labour markets (see Chapter 6).

In Chapter 7, we provide comparisons and an analysis of correlations between the initial ideas and perceptions of the COST Action members and the mutually accepted approaches achieved by EU and PCs universities concerning the curricula for Belarusian and Ukrainian HEIs.

The ERASMUS+ project "CybPhys" includes a number of important innovations. The consortium will create a number of e-books which allow the teaching staff in Belarus and Ukraine to adapt the new and innovative CPS topics to the needs of the local students. This approach includes the creation of a joint web-based e-library with e-books, synopses and additional teaching and learning materials.

The use of e-learning environments will face the technical and educational needs of today and tomorrow. A Sharing Modelling and Simulation Environment (SMSE) platform will be created on shared lab facilities of the partners. Using new computer classes, this SMSE platform enables access to virtual labs.

Learning and teaching methodologies and pedagogical approaches, including learning outcomes and ICT-based practices, will support CPS modelling education. By assisting self-learning and by developing blended courses in combination with ICT-based tools like wikis, GitBook, and Prezi, a broad range of educational approaches arise.

11.2. Related work

The need of providing knowledge timely and accurately arises in different settings, starting from industrial needs to train employees in new technologies and ending with universities that have to become more and more flexible in terms of their curricula and course development [33]. Each study program may be considered to concern specific knowledge delivered by means of particular courses, which should react quickly enough to the industrial requirements. The software engineering discipline is developed under the evolution of new technologies and the dramatically increasing scale of applications. The term Industry 4.0 was first used in 2011 [34] by a group of representatives from different fields, such as business, politics and academia. Industry 4.0 mainly aims to bring together information technology and industry and can be defined as a movement forward from automation of electronic and IT systems to smart factories and cyber-physical systems (CPS). CPS are networked interacting systems which consist of physical and digital components. The physical components represent real physical system objects and machine tools. The digital (or virtual) part serves as a cyber-image of the physical object [35].

As far as CPS are the core elements of Industry 4.0 the market of these systems is rapidly expanding, because of which there arises a rapid rate of changes in the education of specialists in computer science and engineering domains. Interdisciplinary skillset to introduce, model, design, develop and deploy CPS becomes more and more necessary. Many activities have been noticed dealing with issues of cyber-physical systems curricula and education. A brief overview of study programs in highly recognised world universities shows that this exciting technological revolution affected also the content of system engineering education. A lot of on-line courses, election courses and mandatory courses concerning studying different aspects of CPS are offered at different levels of degree. For example, see the courses found in [36], [37].

Moreover, several universities offer a study of CPS as a concentration of system engineering study programs, for example, [38], [39], [40], [41], [42]. Some universities give CPS to be studied as a part of an embedded system program [43], [44], or a mechatronic system program [45], [46], where a block of interrelated courses is given for students to study separately containing a mandatory list of courses, like Introduction to CPS, Design of CPS, Embedded Networking, Cyber Security Practice, and CPS Modelling.

One more interesting finding when looking for CPS study programs available on Internet for application is a Master of Engineering Degree Program in Cyber-Physical Systems, offered by Vanderbilt University. The university itself announces that it "is a global leader in CPS research and education and is in a unique position to offer a well-recognised interdisciplinary graduate program in this area" [47]. This study program is not concentrated as holistic and integral education of the students, but it has flexible academic curriculum, is realised by multiple departments and allows students to tailor their studies to their professional interests and goals by selecting courses from different concentrations.

Despite CPS courses being not offered systematically or being defined as a part of other study programs, both researchers and educators have intensively been publishing their findings on CPS curriculum development the last decade. The authors of [48], [49], [50], [51] discuss the fundamental things which should be included into CPS curricula. Meanwhile, the authors of [52], [53] present a principally new view on the education of system engineers in the aspect of CPS offering the structure and frameworks for practical laboratories and encouraging the principle of "learning by doing". One more example of research focuses on the competences of cyber-physical systems engineers and concludes that specialists in CPS have to acquire an essential interdisciplinary engineering qualification combining skills and competencies from different disciplines, including social and psychological aspects, among others. Therefore, in order to develop and implement an adequate study program, two more major aspects should be taken into consideration: (i) identification and understanding of the minimal set of multidisciplinary competencies and qualifications needed and (ii) identification and classification of the critical gaps in the existing engineering curricula [71], [54]. The authors of [9] and [55] performed similar research with the development of a CPS study program based on a survey about skills and competences required for CPS engineers.

ARTEMIS [12] has founded an Education and Training (E&T) Working Group [13] to establish a strategic agenda for E&T, taking into account the needs of the European industry. The E&T WG wants to achieve a self-sustaining Innovation Environment for European leadership in Embedded Systems (ES) and CPS. A very detailed and comprehensive analysis of the EU policy in Electronic Components and Systems for European Leadership and especially about CPS in the last decade is provided in [14], [15].

To prepare students to meet the required industry competencies [16] and to gain sufficient knowledge in various areas related to CPS, curricula developers must consider also (1) the complex working environments; (2) recent technologies; and (3) tools and skills, in order to prepare their students to fulfil the expectations of the industry [17].

For the education of software engineers, the Systems Engineering body of knowledge (SWEBOK) [18] can help in developing state-of-the-art System Engineering (SE) curricula. It is vital to face the market demands and to align the curricula with the SE models and standards. This includes the SE body of knowledge (SWEBOK) [18], IEEE-CS and ACM SE2004 guidelines [19], IEEE-CS and ACM SE2014 guidelines [20] and the GSwE2009 guidelines (which are particularly for graduate-level curricula) [21]. There is no such body of knowledge for CPS engineering as we have for software engineering, one of the possibilities to define the core knowledge of the CPS expert is using the research results in Ref. [22].

The development of machine to machine learning in Embedded Systems, as well as the steps towards the CPS Approach [24], requires the future development of study programs in the areas of cross integration of computer science, electronics, electrical engineering and telecommunications. The issues of CPS challenges for automation are also depicted in [11], [23], [25].

Practical implementation of courses on modelling and simulation for CPS is discussed in [26], [27], [28]. In the work [26], the authors have developed and offered a course on CPS that relies on a simple high level modelling and simulation language embodied in an interactive environment supporting 3D visualisation. A different approach is suggested in Ref. [27]. This paper proposes a method to infuse CPS concepts into the computer science curriculum via a set of CPS course modules that emphasise modelling and verification. The course proposed in [28] combines the theoretical foundation of modelling with a wide spectrum of modelling and simulation formalisms.

Both [27] and [28] offer advanced curricula for IT program students. However, we pursued a curriculum for students having different backgrounds, e.g. in the different topics of applied physics, electrical engineering, biomedicine, and transport. The approach discussed in [26] looked more appropriate but we had to take into consideration the particular labour market needs of Belarus and Ukraine.

Ref. [31] discusses the implementation of the joint innovative double degree master program (DDMP) in the field of energy-saving technologies (EST) in transport between European and Ukrainian universities. At the same time, the main goals when implementing the DDMP are: to improve higher education in accordance with the changing needs of economic and social spheres in the road transport sector due to the growth of innovative energy-efficient and EST; to increase the competitiveness of graduates in employment and the productive cooperation between the universities; to reduce energy consumption and to replace traditional sources by "green" forms of energy in the transport sector as well.

As in several articles, i.e. [56], [57], [58], [59], [60], [61], research evolution of system engineer curricula shows the aspect of CPS integration within them. Some authors analysed modern trends in CPS industry to conclude to their impact and define requirements of CPS engineers [62], [63], [64], [65], [66]. Overall it is possible to find articles where the authors prove or stress the necessity of the integration of CPS courses to make software engineer's study programs all rounded, comprehensive and complete. Examples or case studies demonstrate how to implement such integrations; see [67], [68], [69], [70].

So far, there is no doubt that the integration of study courses devoted to CPS is vitally important in the curricula of system engineers to match to the modern trends in computer systems. Moreover, the content of this separate stream focused on CPS, in education usually called "concentration", should be based not only on fundamentals of CPS defined by academicians, but also on industrial needs. The next section offers a review of the labour markets in Belarus and Ukraine to show how the review results can affect the development of a CPS study program taking into consideration not only fundamentals, competencies and skills of a CPS engineer, but also specific situational needs required in specific countries.

11.3. Review of the labour markets of Belarus and Ukraine

CPS are large complex systems, where physical elements (and even their systems) interact with and are controlled by a large number of distributed and networked computing elements and human users. The more highly developed the state, the more complex the infrastructure, which ensures its normal functioning, including the well-being of its citizens. To provide this infrastructure, CPS and their systems are required, which ensure efficient operation of both the economy as a whole and its individual elements: all types of industry, transport, energy, manufacturing, security systems, logistics networks, smart houses, etc. The global market of Digital Technology is estimated at EUR 3300 billion, corresponding to around 50 million jobs. Europe's position is characterised by a strong presence in software and services, representing 8.9 million jobs (Multi Annual Strategic Research and Innovation Agenda for the ECSEL Joint Undertaking (MASRIA 2017)).

11.3.1 The needs of the Belarusian labour market

Belarus in this sense is a fairly developed country in which there is a majority of the above systems, which can be attributed to the CPS. Therefore, the organisation of effective and well-coordinated work of this kind of CPS requires the provision of their respective monitoring and control systems, in which the information and communication technology (ICT) plays a major, if not decisive, role. The problem of creating and using such systems is multidisciplinary, since it requires the collaboration of tooling and solution providers, end-users, and research institutions, which must create a CPS and seek to towards the development of cognitive cyber-physical systems. To plan and direct the stable improving evolution of Belarusian economy the specially trained staff with specific skills and competences is required, which should be capable to develop, design and control such systems. In this connection, Belarus is faced with very important and very pressing problems — creating an improved education system that would prepare both consumers of already known solutions in this field and developers of new (future) CPS control and monitoring. In this sense, the integration of classical, technical and pedagogical universities in one project aimed at developing a modern educational environment in the field of CPS modelling is very timely, promising and correct. To create an effective system for education in the mentioned problematics, Belarusian teachers, researchers and engineers need to use experience of their colleagues from EU countries universities.

Analysis of the labour market in Belarus shows that graduates of the Belarusian State University (BSU), Gomel State University (GSU) and Mozyr State Pedagogical University (MSPU) are to be trained in the new specialties and some of profiles developed under the project will be in demand firstly by ICT high-tech companies and design centres belonging to the High-Tech Park and producing companies and also their Research/Educational Institutions in Belarus and Ukraine.

11.3.2 The needs of the Ukrainian labour market

The constantly growing number of hybrid and electric vehicles requiring a completely new approach to the development of infrastructure for maintaining and repair work proves this fact. This approach should be based on the implementation of contemporary, innovative energy-saving (ES) and energy-efficient (EE) systems.

Monitoring of existing educational programs for Master's degrees in motor vehicles specialties has shown that relevant knowledge and skills of ES and EE technologies in transport and electric vehicles are actually not provided by any technical educational institution in Ukraine. However, several EU universities have already studied new master programs related to the exploitation and maintenance of hybrid- and electro transport vehicles, e.g., Silesian University of Technology – "Eco- & Electro Systems of Auto Vehicle"; Warsaw University of Technology – "Electric and Hybrid Vehicle Engineering", etc.

The Eastern region of Ukraine is a centre of metallurgical and mining industries and the universities take part in research activities concerning energy saving, environment protection and ecology monitoring.

The current training of specialists in the field of energy saving technologies covers a very broad range of general knowledge. Therefore, this way of training for highly specialised workers makes such knowledge very impractical and hard to apply in present-day companies' operation which use recent EE technologies in transport and are engaged in the manufacture and/or sale and/or maintenance of hybrid and electric vehicles (EV). The new training programs of Kharkiv National Automobile and Highway University (KhNAHU) and Kryvyj Ryg National University (KNU) have to accumulate the vital range of disciplines and knowledge to work in such companies. To plan and implement the stable development of EE road transport infrastructure, specially trained staff is required, who should be capable to develop and design such systems, as well as carry out the necessary maintenance and repair work of not only hybrid and electric vehicles but also of electric, electronic and information systems of this infrastructure.

The implementation into the educational process of such joint (for all project participants) educational programs will result in standard (double, multilateral) Master's diplomas in EV & EST between KhNAHU and RTU.

The draft National Strategy of Industry 4.0 in Ukraine, issued in December 2018, states that for the domestic market the Industry 4.0 should become a catalyst for the growth of industry as well as for defence. The specialists in industrial electronics have been trained at the Chernigov National University of Technologies (CNTU) for more than 30 years. As of lately the labour market requires specialists with competences at Industry 4.0. This could be achieved through the introduction of a new master program in Industrial Automation (IA). The process of learning an IA should be based on modelling techniques provided by laboratory equipment with specific software. The teaching of IA requires a good mathematical preparation and the knowledge of recent advances in simulation of cyber-physics systems, model-oriented control, design of electronic components. Involving of EU partners, which have experience at the training in IA would reinforce implementation of such an education program. Internships for specialists will be organised in companies of the Chernihiv IT-cluster, and regional enterprises that produce automation lines.

11.4. COST Action input to European curricula on CPS

COST Action Multi-Paradigm Modelling for CPS (MPM4CPS) was implemented in the years 2014–2018 with primary scientific targets: the conceptualisation of techniques and tools for improving interoperability; the development of new ontologies and formalisms to deal with the heterogeneity; and, to perform the integration of the problems resulting from several application domains, under a common MPM4CPS umbrella [5]. This Action also aimed to create the base for a European Master and a PhD program in MPM4CPS. The Action aimed to set up the respective discipline roadmap for the development of a mutually recognised cross-domain expertise-based study program in CPS. RTU was one of the Action members.

Additionally, the Action pursued to develop CPS expert profiles into a suitable format for educational purposes. It has been performed in several steps; one of them is a survey among the partners in order to identify and classify the competencies, which partners recognise as indispensable and most relevant related to particular areas of CPS.

Within the scope of the COST Action, the team conducted a survey among the members in order to identify and classify the competencies, which they recognise most relevant related to particular areas of CPS. These competencies are applicable in academic educational programs [9]. The survey included a set of questions where participants were asked to rate, according to their perception, the relevance of each of the following topics for MPM4CPS, using a Likert scale from 1 (not relevant) to 5 (very relevant). The results presented an overview of descriptive statistics (the mean scores for the classification of relevance for each category).

The results were analysed with two different degrees of granularity. At a higher abstraction level (first level topics, in the list of potential MPM4CPS subjects proposed in the survey structure), we had an overview of the most relevant broad topics. At a finer granularity level, we also observe differences in the distribution of the classifications provided by our participants to the finer-grained topics. For example, the sub-topics directly concerning CPS have a mode close to five (very relevant) with the mean 4.5385. Sub-topics concerning MPM, Application Domains and Software Engineering were considered relevant i.e. close to 4. Finally, design and simulation have a mode of 3 (the neutral value), with the vast majority of answers provided between 3 (the mode) and "relevant". The different numbers of answers for some of the topics at a finer granularity level revealed that not all participants were familiar enough with all the topics to feel comfortable in answering about each topic's adequacy [9].

Overall, both for the higher abstraction level and for the fine-grained topics, the vast majority of opinions ranged from very positive to neutral, with few exceptions. We considered such kind of ranking as a relevant element when discussing a CPS curriculum, but not as a defining metric for including or excluding a particular topic from the CPS curriculum.

The results of the survey have been utilised in two areas. First, the method of the survey arranged in COST Action was successfully applied for a survey among Belarusian and Ukrainian prospective employers (see Chapter 5). Secondly, results of the survey were used as references, when we discussed the topics of curricular for Belarusian and Ukrainian partners. Due to lack of space we do not provide more data, please see [9].

11.5. Identification of industry needs

11.5.1 Method of the research on curricular in CPS

In 2018 Riga Technical University (Latvia), University of Cyprus, KU Leuven University (Belgium), in collaboration with six Belarusian universities and two professional associations successfully completed "Improvement of master-level education in the field of physical sciences in Belarusian universities" (Physics), the project awarded under the Erasmus+ CBHE [10]. Valuable experience obtained in collaboration between three EU and six Belarusian universities encouraged one to expand further collaboration by including in consortium three Ukrainian universities. Therefore, a new ERASMUS+ project "Development of practically-oriented student-centred education in the field of modelling of Cyber-Physical Systems", "CybPhys", started in 2019, to be implemented by a consortium comprising three EU universities (RTU, KU Leuven, UCY), three Belarusian universities (BSU, GSU, MSPU), RANI (Association of Nano-industry Enterprises) and three Ukrainian universities (KhNAHU, KNU and CNTU).

The wide project goal is to upgrade bachelor and master-level curricula and study programs according to Bologna practices in Belarusian and Ukrainian universities in the area of Cyber-Physical Systems modelling and simulation. These curricula are directed on innovative physical, mathematical and engineering sciences and High-Tech industry topics. An important focus of the project is the use of ICT through networking activities to the labour market needs and enhancement of the quality and relevance of education. The specific project objectives are focused on the further reformation of PCs HEIs according to the ET2020 strategy of European Union:

- To modernise the bachelor- and master-level curricula and study programs for the Physical, Mathematical, and Engineering Faculties in six universities of Belarus and Ukraine according to EU university's practices in the area of innovative modelling and simulation of CPS for High-Tech industry and scientific research institutions;
- to enhance the quality of education in the area of modelling and simulation of CPS, based on the modernised bachelor and master-level training programs, focusing on the use of innovative ICT environment to realise the declared targets;
- To supply relevance of higher education in the area of modelling and simulation of CPS in Belarus and Ukraine to the main instruments and principles of Bologna process, and such European Higher Education Area (EHEA) documents, as ISCED 2011, a Framework for Qualifications of the EHEA, ECTS, Standards for quality assurance in the EHEA, etc.

Being one of the partners of COST Action MPM4CPS and the coordinator of ERASMUS+ project, RTU together with the project partners recognised that a synergy of COST and ERASMUS+ projects provides an excellent opportunity for the development and validation of labour market-oriented training programs. Therefore, the findings and recommendations derived from COST Action may be approbated and validated at HEIs of Belarus and Ukraine.

The offered methodology aims to develop and validate the CPS expert's profile suitable for Belarusian and Ukrainian labour market needs. It should be done in four steps:

- Appreciation methodology, techniques and deliverables created in COST Action;
- analysis of advances on CPS curricular and mapping against the needs of the labour markets in Belarus and Ukraine;
- validation of CPS specialist profiles in a frame of a CybPhys project, which the partners will implement in years 2019–2022;
- involving of Belarusian and Ukrainian stakeholders in industry, science and HEIs to the improvement of education in the field of CPS modelling through developing/modernising new types of training programs.

The project partners have got valuable experience on the investigation of the needs of the labour market, for example, in 2018, in a frame of preparation of application form for future ERASMUS+ CybPhys project, a pre-survey of professional associations, research institutes, and universities as employers of master graduates was arranged. It was aimed on the investigation of the specific needs of the labour market in Belarus and Ukraine in the field of CPS modelling when realising a transition to Bologna principles in education (see, below).

The CybPhys project focuses on a curricula modernisation taking into account the results of the labour market need analysis, and forecast of its development in close future, taking into account recommendations of non-government bodies, design centres, research institutions and HEIs in Belarus and Ukraine. For this purpose, in December 2018–January 2019, we arranged a preliminary survey of research institutions, professional associations, companies and HEIs of Belarus and Ukraine. Additionally, we arranged a similar survey in Belarus in November 2017.

Findings and data obtained from this survey are intended to be used for drafting future models and curricula for undergraduates and master students with a two-year training cycle in the field of applied physics and engineering specialties. The questions of the survey are devoted to the needs concerning application topics in modelling of CPS.

We adjusted a polling method and a questionnaire applied in COST MPM4CPS taking into account national particularities of research institutions of Belarus. We evaluated the relevance of each of the possible topics for training of specialists in the field of CPS in accordance with respondent' perception. A Likert scale from 1 (not applicable) to 5 (very important) was applied.

11.5.2 Analysis of survey of research institutions of Belarus

15 respondents participated in the polling: from Research institutes of NAS of Belarus – 5, HEIs and their Research Institutes – 8, Private Enterprises – 2. We also wanted to know the extent to which the respondents had prior experience with CPS industrial projects and modelling of CPS. Experience in work on the projects related

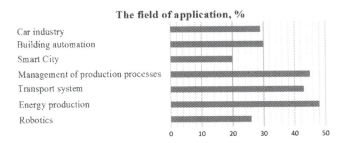

Figure 11.1 Seven high level possible topics for training of specialists in the field of CPS.

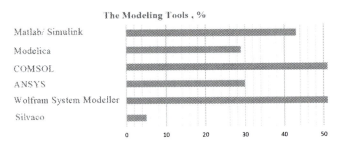

Figure 11.2 The histogram concerning questionnaire respondents evaluated the relevance of industrial design tools.

to computer modelling of physical processes and CPS denoted 11 persons of 15. To the question about experience in modelling of physical processes and CPS all 8 persons answered "yes", but 5 persons accepted the lack of experience.

There were identified seven high level possible topics for training of specialists in the field of CPS in accordance with the respondents' perception, presented in Fig. 11.1.

In the Design and Simulation part of the questionnaire, the responders evaluated the relevance of industrial design tools (see Fig. 11.2). Take note that a majority of responders got a preference in usage of COMSOL and Wolfram System Modeler.

Due to the lack of space in this paper, we restricted the analysis of responses to only these examples.

11.5.3 Analysis of survey of enterprises of Ukraine

From the Ukrainian side, 24 respondents participated in the survey, their professional activities included: transport, maintenance, and repair of vehicles, cargo delivery – 14; education and research in the field of automobile transport – 7; Software Development – 2; electrical equipment – 1.

Experience in work on the projects related to computer modelling of physical processes and CPS denoted 10 persons of 24. Experience in modelling cyber-physical systems made up 6 persons of 24.

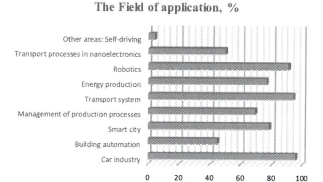

Figure 11.3 Nine high level possible topics for training of specialists in the field of CPS.

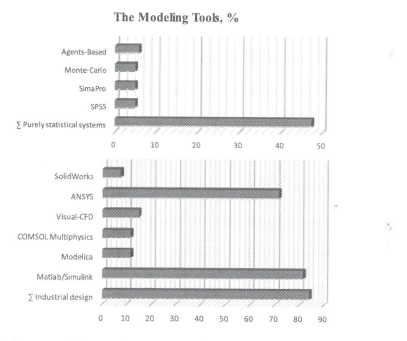

Figure 11.4 The histogram of the questionnaire respondents about evaluation of modelling tools.

There were identified nine high level possible topics for training of specialists in the field of CPS in accordance with the respondents' perception, presented in Fig. 11.3.

In the Design and Simulation part of the questionnaire responders preferred Industrial design tools with the almost double advantage in comparison with Purely statistical systems, presented in Fig. 11.4.

It should also be noted that in the part of "Software development" respondents were most interested in "Resource-intensive modelling" and "Autopilot, control system". In the part of "Other relevant areas of student training", responders' interest focused on: "Autonomous transport (electric transport)"; "Smart roads"; "Transport infrastructure". Therefore, in the training courses developing within the framework of ERASMUS+ project "CybPhys" for Ukrainian universities, it is necessary to take into account the obtained data.

11.5.4 Findings of survey of research institutions

Preliminary results of the survey revealed, with areas of CPS applications, modelling methods, tools and languages, etc., can be considered as more preferable for teaching graduates looking for a job in research companies. Actually, the fact that one third of the participants provided a negative answer to the question "Do you have experience in modelling of physical processes and CPS" gives an idea to evaluate the answers with caution, due to the small number of experienced respondents. However, this fact shows the real situation related experience in CPS.

We also found similarities in the evaluation provided by EU and Belarusian respondents. For example, ranking the modelling tools by European, Belarusian and Ukrainian respondents put higher marks to Modelica and MATLAB® than SPSS and SigmaPro.

According to the educational field specifics of Ukrainian universities, there is an interest in the field of automobile transport and its infrastructure (autopilot, control systems, smart roads, etc.). The use of CPS tools and methods in the educational field, which are provided by the ERASMUS+ CybPhys project, will give a significant increase both in the quality of education and in the creation of highly qualified specialists in this area. This, in turn, meets the ever-increasing current requirements of the labour market.

11.6. Validation of COST findings for curricula on CPS in Belarusian and Ukrainian universities

Recently, the team of 6 universities of PC and 3 of EU countries started to work in the framework of new ERASMUS+ project "CybPhys" aiming to develop the Industry Oriented Curricular on CPS for Belarusian and Ukrainian universities. The target of this project is to develop new curricula and didactic tools (courses, virtual practices, electronic manuals and books), to improve the training of bachelor- and master-level students in the field of computer modelling and simulation, to accelerate the transfer of knowledge and new educational technologies in this area (including ICT tools), to enhance the exchange of experiences in joint training of scientific personnel, to train specialists with new skills and competences, etc. In order to increase synergy of the partners, we suggested:

- To develop the system of interconnected didactic tools for teaching/learning of students: 9 e-books (see Table 11.1) and appropriated courses (see Table 11.2).

Table 11.1 The list of developed e-books.

N	E-book names
1	Bringing innovations to the market
2	Mathematical Modelling of Mechatronic Systems
3	Model-oriented control in Intelligent Manufacturing Systems
4	Modern Mathematical Physics: Fundamentals and Application
5	High-Performance Scientific Computing and Data Analysis
6	Cyber-Physical Systems modelling and simulation
7	Cyber-Physical Systems for Clean Transportation
8	Control methods for critical infrastructures and Internet of Things (IoT) interdependencies analysis
9	Computer modelling of physical processes (handbook for students and PhD students)

- To consolidate theoretical knowledge and practical attainments obtained by students in the interrelated courses with the support of a system of virtual laboratory workshops created on the basis of purchased hardware and software systems and newly developed didactic tools.
- To use jointly developed courses and workshops in partners' universities through the use of trainings in the framework of the Sharing Modelling and Simulation Environment (SMSE) platform developed in Belarusian State University and distance e-learning (DEL) methods.

Therefore, the implementation of these approaches, SMSE and DEL will allow one to create joint educational system for teaching/learning of bachelor- and master-level students by computer modelling techniques as a system of compatible curricula, didactic (e-books, courses, virtual workshops, instructive guides) and technical (Moodle platform, interactive smartBoards, GoogleJam screens) tools.

A set of developed e-books, courses, workshops and guides will provide training in the field of both basic and special disciplines of high complexity. This will ensure the consistent training of students for the entire Bachelor-Master-PhD chain in the targeted areas of CPS modelling/simulating. The offered system of e-books, courses and virtual labs can be conditionally divided into the following four blocks:

- To prepare a specialist in engineering and physics for the basics of modern data analysis and calculation a course e-book will be developed for basic preparation at the bachelor's level (e.g., book 4 "Modern Mathematical Physics: Fundamentals and Application"), providing a strong basis for future more specialised courses. It can be regarded as a prerequisite for other books of special and advanced training.
- e-Book 5 "High-performance computing in numerical simulations and data analysis" will provide users with specialised knowledge for working with large data and simulations using supercomputers. This lays the groundwork for an in-depth

Table 11.2 The list of courses connected with the blocks.

Numbers of e-Book in Table 11.1	Course names
1	Fundamentals of business and legislation in IT, Bringing innovation to the market, Entrepreneurship in the ICT sector, Project management, E-commerce.
2	Dynamics and Stability of Mechanical Systems, Theoretical Mechanics, Analytical modelling of friction and wear processes Computer modelling, Analytical modelling of friction and wear processes.
3	Model-oriented control in Digital Manufacturing, Programming of Automation Systems, Design and Simulation of Power electronics components, Modelling and Measurement of physical processes in Robotics, Simulation of Manufacturing Environment
4	Equations of mathematical physics, Fundamentals of mathematical modelling, Methods of mathematical physics, Mathematical modelling of physical processes.
5	Supercomputer Programming, High-Performance computing for numerical simulations and data analysis, Mathematical modelling for fluid- and gas dynamics, Data mining and acquisition, Receiving and in-depth analysis of data, etc.
6	Big data and the need for data processing, Cyber-security, Power systems, SCADA system, smart grid, simulation and modelling, wide area monitoring and control, Communication systems and tools for cyber-physical systems, etc.
7	Energy-saving technologies in transport, The structure of hybrid and electric vehicles, Electric systems of environmentally friendly vehicles, Methods of planning scientific research on vehicles, Mathematical modelling and optimisation methods, etc.
8	Applications of computer modelling, Internet of Things, Operating Systems, etc.
9	Computer modelling of physical systems, processes and phenomena, Modern integrated packages for analysis and modelling of processes and systems, Object oriented programming, MATLAB, Simulink and Flash for solution and computer simulation of research problems in Physics, etc.

understanding of advanced computing technologies. Based on this book of special training, an advanced course will be built as a SPOC (Small Private Online Course), which will be used in the training at master-level and PhD-level.

- Courses based on e-books 2–3 and 6–9 for special training will help students to specialise in different areas of modelling (depending on market requirements and training profiles in the respective universities). The topics are selected from the fastest growing modern sections of the industry, as well as applied and fundamental physics.

- e-Book 1 "Bringing innovations to the market" allows students to fully discover their creative potential by providing an understanding of the mechanisms of implementations in innovative entrepreneurship in high-tech industries. Strengthening the orientation towards start-up projects is also a goal.

11.7. Discussions and conclusions

Cooperation between COST and ERASMUS+ project teams provided benefits to both projects. The COST team shared its experience concerning analysis of tendencies, industry needs, and acquiring the best educational practice. The approaches developed in the COST project have been tested and validated in the ERASMUS+ project Physics and a similar approach will be practised in the new ERASMUS+ project CybPhys.

There are important similarities in the answers provided by European, Belarusian and Ukrainian respondents. For example, the ranking of the modelling tools provided by the respondents in the EU and the respondents in the so-called Partner Countries was rather similar. However, when comparing the answers given in the EU and the partner countries, it is important to bear in mind the different experiences and competences of the respondents. It is also important to bear in mind the different socio-economic environments of the EU and the partner countries.

The analyses of the answers revealed a very high interest of the employers concerning the knowledge and the skills of graduates, required for development and application of CPS. There are differences between the answers of Belarusian and Ukrainian respondents: all Belarusian respondents had experience in modelling of CPS; however, only one fourth of the Ukrainian respondents had such experience. The Belarusian respondents mostly represented research institutions of the Academy of Science, while the Ukrainian respondents mostly represented transport, logistic and mining industrial companies. The Belarusian respondents are mainly interested in the modelling of CPS used to control different processes in the fields of applied physics. The Ukrainian respondents are mainly interested in knowledge of CPS used to control processes in transportation, energy production and manufacturing.

The consortium is aware of restrictions of the survey on the topics of modelling of CPS for new curricula due to the limited number of the respondents in Belarus and Ukraine. Therefore, an additional survey is planned at the initial stage of the CybPhys project. The ERASMUS+ project CybPhys will bring novelty to the higher educational systems of Belarus and Ukraine by developing e-books, an e-library, e-learning environments, a Sharing Modelling and Simulation Environment platform and the CPS oriented courses.

Despite the different backgrounds of the respondents and differences between the labour markets, the inspiring principles of the Bologna process are useful in combination with the introduction of self-learning stimuli, blended courses and ICT-based learning and teaching tools.

Acknowledgements

This publication was developed with the support of COST Action IC1404 "Multi-Paradigm Modelling for Cyber-Physical Systems", ERASMUS+ project "Improvement of master-level education in the field of physical sciences in Belarusian universities" (Physics) and ERASMUS+ project "Development of practically-oriented student-centred education in the field of modelling of Cyber-Physical Systems" (CybPhys).

References

[1] A. Hnatov, S. Arhun, K. Tarasov, H. Hnatova, V. Mygal, A. Patlins, Researching the model of electric propulsion system for bus using Matlab Simulink, in: USB PROCEEDINGS of 2019 IEEE 60th International Scientific Conference on Power and Electrical Engineering of Riga Technical University (RTUCON), Latvia, Riga, 7–9 October, 2019, Riga Technical University, Riga, Latvia, ISBN 978-1-5386-6902-0, 2019, pp. #051-1–#051-6.

[2] Yetis Hasan, Mehmet Baygin, Mehmet Karakose, An investigation for benefits of cyber-physical systems in higher education courses, in: 15th International Conference on Information Technology Based Higher Education and Training (ITHET), IEEE, 2016, https://doi.org/10.1109/ITHET.2016.7760734.

[3] J. Peuteman, A. Janssens, J. Boydens, D. Pissoort, Integrating research and design oriented skills in a learning trajectory for engineering students, in: Proceedings of the EDULEARN18 Conference, Palma, Mallorca, Spain, July 2–4, 2018, pp. 877–886.

[4] L. Gitelman, M. Kozhevnikov, O. Ryzhuk, Advance management education for power-engineering and industry of the future, Sustainability 11 (21) (2019) 5930 (pp. 1–22), https://doi.org/10.3390/su11215930.

[5] Multi-Paradigm Modelling for Cyber-Physical Systems (MPM4CPS) COST Action IC1404, http://www.cost.eu/COST_Actions/ict/IC1404?parties.

[6] ERASMUS+ CBHE Capacity-building in the Field of Higher Education 2015 Call for Proposals Improvement of master-level education in the field of physical sciences in Belarusian universities, (Physics), http://physics.rtu.lv/. (Accessed November 2017).

[7] J. Peuteman, A. Janssens, R. De Craemer, J. Boydens, A. Zabasta, A. Fedotov, Integration of the European bachelor master degree concept at Belarusian universities for physics and engineering students, in: Proceedings of the XXVth International Conference Electronics – ET, Sozopol, Bulgaria, September 12–14, 2016.

[8] A. Zabašta, N. Kuņicina, J. Peuteman, R. De Craemer, A. Fedotov, Development of industry-oriented, student-centred master-level education in the field of physical sciences in Belarus, in: Proceedings of EDULEARN18 Conference, Palma, Mallorca, Spain, 2–4 July, 2018, pp. 3641–3648.

[9] A. Zabašta, P. Carreira, O. Nikiforova, V. Amaral, N. Kuņicina, M. Goulão, U. Sukovskis, L. Ribickis, Developing a mutually-recognized cross-domain study program in cyber-physical systems, in: 2017 IEEE Global Engineering Education Conference (EDUCON 2017), Greece, Athens, 25–28 April, 2017, IEEE, Piscataway, 2017, pp. 791–799, available from: https://doi.org/10.1109/EDUCON.2017.7942937.

[10] S. Fedotov, A. Fedotov, A. Tolstik, A. Zabašta, A. Žiravecka, N. Kuņicina, L. Ribickis, Evaluation of market needs in Belarus for improvement of master-level education in the field of physical sciences, in: 2016 57th International Scientific Conference on Power and Electrical Engineering of Riga Technical University (RTUCON), Latvia, Riga, 13–14 October, 2016, IEEE, Piscataway, NJ, 2016, pp. 1–6, https://doi.org/10.1109/RTUCON.2016.7763148.

[11] Master's Program in Pervasive Computing and Communications for Sustainable Development (PERCCOM), http://www.lut.fi/web/en/admissions/masters-studies/msc-in-technology/information-technology/erasmus-mundus-programme-in-pervasive-computing-and-communications-for-sustainable-development.

[12] ARTEMIS Joint Undertaking (Advanced Research and Technology for Embedded Intelligence and Systems), www.artemis-ju.eu, ARTEMIS Strategic Research Agenda (SRA) (2011 and 2013).

[13] Artemis-IA, WG Education & Training, https://artemis-ia.eu/working-groups/wg-education-training.html.

[14] E. Schoitsch, A. Skavhaug, European perspectives on teaching, education and training for dependable embedded and cyber-physical systems, in: 2013 39th Euromicro Conference Series on Software Engineering and Advanced Applications.

[15] E. Schoitsch, Special Session TET-DEC (Teaching, Education and Training for Dependable Embedded and Cyber-Physical Systems), An E&T use case in a European project, in: 2015 41st Euromicro Conference on Software Engineering and Advanced Applications.

[16] H. Jaakkola, J. Henno, I.J. Rudas, IT curriculum as a complex emerging process, in: IEEE International Conference on 2006 ICCC Computational Cybernetics, IEEE, IEEE Society, 2006, pp. 1–5, https://doi.org/10.1109/ICCCYB.2006.305731.

[17] A. Alarifia, M. Zarour, N. Alomar, Z. Alshaikh, M. Alsaleh, SECDEP: software engineering curricula development and evaluation process using SWEBOK, Information and Software Technology 74 (2016) 114–126.

[18] P. Bourque, R.E. Fairley (Eds.), Guide to the Software Engineering Body of Knowledge, Version 3.0, IEEE Computer Society, 2014, www.swebok.org.

[19] R.J. LeBlanc, A. Sobel, J.L. Diaz-Herrera, T.B. Hilburn, et al., Software Engineering 2004: Curriculum Guidelines for Undergraduate Degree Programs in Software Engineering, IEEE Computer Society, 2006.

[20] T.J.T.F. on Computing Curricula, Software Engineering 2014: Curriculum Guidelines for Undergraduate Degree Programs in Software Engineering, Technical Report, New York, NY, USA, 2015.

[21] A. Pyster, Software Engineering 2009 (GSwE2009): Curriculum Guidelines for Graduate Degree Programs in Software Engineering, Integrated Software & Systems Engineering Curriculum Project, Stevens Institute of Technology, 2009.

[22] Interim Report on 21st Century Cyber-Physical Systems Education, available at: https://www.nap.edu/read/21762/chapter/2, 2015.

[23] C. Scaffidi, What training is needed by practicing engineers who create cyberphysical systems?, in: Proceeding of 2015 41st Euromicro Conference on Software Engineering and Advanced Applications, pp. 298–305.

[24] C.A. Berkeley, LeeSeshia.org [Online]. Available: http://LeeSeshia.org, 2011.

[25] D. Vasko, et al., Foundations for Innovation in Cyber-Physical Systems Workshop report, January 2013 Prepared by Energetics Incorporated Columbia, Maryland for the National Institute of Standards and Technology, p. 60.

[26] W. Taha, R. Cartwright, R. Philippsen, Y. Zeng, Developing a first course on cyber-physical systems, in: Workshop on Embedded and Cyber-Physical Systems Education WESE'14, New Delhi, India, October 12–17, 2014, 2014, pp. 1–8, https://doi.org/10.1145/2829957.2829964.

[27] K. Damevski, B. Altayeb, H. Chen, D. Walter, Teaching cyber-physical systems to computer scientists via modeling and verification, in: Proceeding of the 44th ACM Technical Symposium on Computer Science Education SIGCSE'13, Denver, Colorado, USA, March 6–9, 2013, pp. 567–572.

[28] Y.V. Tendeloo, H. Vangheluwe, Teaching the fundamentals of the modelling of cyber-physical systems, in: SpringSim-TMS/DEVS 2016, Pasadena, CA, USA, April 3–6, 2016, Society for Modeling & Simulation International (SCS), 2016, pp. 1–8.

[29] A. Zabasta, N. Kunicina, Y. Prylutskyy, J. Peuteman, A.K. Fedotov, A.S. Fedotov, Development of industry oriented curricular on cyber physical systems for Belarusian and Ukrainian universities, in: Proceedings of the 6th IEEE Workshop on Advances in Information, Electronic, and Electrical Engineering, AIEEE 2018, Vilnius, Lithuania, November 8–10, 2018.

[30] J. Peuteman, A. Janssens, R. De Craemer, H. Hallez, P. Coudeville, C. Cornelly, A. Maricau, A. Degraeve, G. Strypsteen, P. Rauwoens, A. Zabasta, Realizing an international student exchange program for Belarusian engineering students to Belgium, in: Proceedings of the SEFI 2017 Annual Conference, September 18–21, 2017, Azores, Portugal, pp. 1142–1149.

[31] A. Gnatov, S. Argun, O. Ulyanets, Joint innovative double degree master program «energy-saving technologies in transport», in: 2017 IEEE First Ukraine Conference on Electrical and Computer Engineering (UKRCON), IEEE, 2017, pp. 1203–1207.

[32] H. Lasi, P. Fettke, H.G. Kemper, T. Feld, M. Hoffmann, Industry 4.0, Business & Information Systems Engineering 2 (6) (2014) 239–242.

[33] O. Nikiforova, V. Nikulsins, U. Sukovskis, Principles of model driven architecture for the task of study program development, in: Frontiers in Artificial Intelligence and Applications, vol. 155: Databases and Information Systems IV, 2008, pp. 291–304.

[34] Klaus Schwab, The Fourth Industrial Revolution, Encyclopaedia Britannica, https://www.britannica.com/topic/The-Fourth-Industrial-Revolution-2119734.

[35] K. Babris, O. Nikiforova, U. Sukovskis, Brief overview of modelling methods, life-cycle and application domains of cyber-physical systems, Applied Computer Systems 24 (1) (2019) 5–12, https://doi.org/10.2478/acss-2019-0001.

[36] University of California, Santa Cruz, Cyber-Physical Systems: Modeling and Simulation, https://www.coursera.org/learn/cyber-physical-systems-1.

[37] University of Oslo, INF5910CPS – Cyber physical systems, https://www.uio.no/studier/emner/matnat/ifi/nedlagte-emner/INF5910CPS/.

[38] Wayne State University, College of Engineering Cyber-Physical Systems, https://engineering.wayne.edu/cyber/curriculum.php.

[39] Universita Degli Studi Firenze, Curriculum "Resilient and Secure Cyber Physical Systems", https://www.informaticamagistrale.unifi.it/vp-153-curriculum-cyber-physical-systems.html?newlang=eng.

[40] Uni Freiburg, Concentration: Cyber-Physical Systems (CPS), https://www.informatik.uni-freiburg.de/studies/furtherinformation/concentrationCPS.

[41] DePaul, College of Computing and digital media, Bachelor of Science, Cyber Physical Systems Engineering, https://www.cdm.depaul.edu/academics/Pages/BS-in-Cyber-Physical-Systems-Engineering.aspx.

[42] Indiana University, School of Informatics, Computing, and Engineering Bulletin, BS in Intelligent Systems Engineering, Computer Engineering/Cyber-Physical Systems Concentration, https://bulletin.iu.edu/iub/soic/2018-2019/undergraduate/degree-programs/bs-intellgient-systems-engineering/computer-engineering-cyber-physical-systems.shtml.

[43] University of Technology Eindhoven, Master Embedded Systems, Cyber-Physical Systems stream, https://educationguide.tue.nl/programs/graduate-school/masters-programs/embedded-systems/curriculum/cyber-physical-systems-stream/.

[44] University of California, Embedded and Cyber-Physical Systems, MECPS, https://www.mastersportal.com/studies/291499/embedded-and-cyber-physical-systems.html.

[45] German Academic Exchange Service, Mechatronic and Cyber-Physical Systems (MEng), https://www2.daad.de/deutschland/studienangebote/international-programmes/en/detail/5378/.

[46] Deggendorf Institute of Technology, Master Mechatronic and Cyber-Physical Systems, https://www.th-deg.de/en/tc-cham-en/courses/master-mechatronic-and-cyber-physical-systems.

[47] Vanderbilt University, Master of Engineering Degree Program in Cyber-Physical Systems, https://engineering.vanderbilt.edu/academics/m_eng/CPS/index.php.

[48] C. Zintgraff, C.W. Green, J.N. Carbone, A regional and transdisciplinary approach to educating secondary and college students in cyber-physical systems, in: Applied Cyber-Physical Systems, Springer, New York, 2014, https://doi.org/10.1007/978-1-4614-7336-7_3.

[49] M. Törngren, et al., Education and training challenges in the era of cyber-physical systems: beyond traditional engineering, in: Proc. WESE'15, Workshop on Embedded and Cyber-Physical Systems Education, Amsterdam, October 8, 2015, 2015, Paper 8, https://doi.org/10.1145/2832920.2832928.

[50] W. Grega, A.J. Kornecki, Real-time cyber-physical systems: transatlantic engineering curricula framework, in: Proc. FedCSIS'2015, Federated Conference on Computer Science and Information Systems, Lodz, Poland, September 13–16, 2015, 2015, pp. 755–762, https://doi.org/10.15439/2015F45.

[51] J. Wade, et al., Systems engineering of cyber-physical systems: an integrated education program, in: Proc. ASEE'2016, 123rd Annual ASEE Conference, New Orleans, LA, June 26–29, 2016, Paper No. 17162.

[52] Janusz Zalewski, Fernando Gonzalez, Evolution in the education of software engineers: online course on cyberphysical systems with remote access to robotic devices, International Journal of Online and Biomedical Engineering (iJOE) 13 (8) (2017), https://doi.org/10.3991/ijoe.v13i08.7377.

[53] S. Peter, F. Momtaz, T. Givargis, From the browser to remote physical lab: programming cyber-physical systems, in: Proc. FIE'2015, Frontiers in Education Conference, El Paso, Texas, October 21–24, 2015.

[54] Elena Mäkiö-Marusik, Bilal Ahmad, Robert Harrison, Juho Mäkiö, Armando Walter Colombo, Competences of cyber physical systems engineers — survey results, in: Industrial Cyber-Physical Systems (ICPS), IEEE, 2018.

[55] N. Kuņicina, A. Zabašta, O. Ņikiforova, A. Romānovs, A. Patļins, Modern tools of career development and motivation of students in Electrical Engineering Education, in: Proceedings of 2018 IEEE 59th International Scientific Conference on Power and Electrical Engineering of Riga Technical University (RTUCON), Latvia, Riga, Nov 12–14, 2018, IEEE, Piscataway, 2018, pp. 1–6.

[56] D. Mourtzis, E. Vlachou, G. Dimitrakopoulos, V. Zogopoulos, Cyber-physical systems and education 4.0 – the teaching factory 4.0 concept, in: 8th Conference on Learning Factories 2018 – Advanced Engineering Education & Training for Manufacturing Innovations, https://doi.org/10.1016/j.promfg.2018.04.005.

[57] Abul K.M. Azad, Reza Hashemian, Cyber-physical systems in STEM disciplines, in: SAI Computing Conference 2016, London, UK, July 13–15, 2016, pp. 868–884.

[58] Isam Ishaq, Rashid Jayousi, Salaheddin Odeh, et al., Work in progress – establishing a master program in cyber physical systems: basic findings and future perspectives, in: 2019 International Conference on Promising Electronic Technologies (ICPET), pp. 4–10.

[59] Martin Törngren, Martin Edin Grimheden, Jonas Gustafsson, Wolfgang Birk, Strategies and considerations in shaping Cyber-Physical Systems education, SIGBED Review 14 (1) (October 2016) 53–60.

[60] Martin Edin Grimheden, Martin Törngren, Towards curricula for cyber-physical systems, in: WESE'14, New Delhi, India, October 12–17, 2014, https://doi.org/10.1145/2829957.2829965.

[61] Linda Laird, Strengthening the "engineering" in software engineering education: a software engineering bachelor of engineering program for the 21st century, in: 2016 IEEE 29th International Conference on Software Engineering Education and Training, pp. 128–131.

[62] Thi Bich Lieu Tran, Martin Törngren, Huu Duc Nguyen, Radoslav Paulen, Nancy Webster Gleason, Trong HaiDuong, Trends in preparing cyber-physical systems engineers, Cyber-Physical Systems 5 (2) (2019) 65–91, https://doi.org/10.1080/23335777.2019.1600034.

[63] A 21st Century Cyber-Physical Systems Education, report, Committee on 21st Century Cyber-Physical Systems Education, Nat'l Academies Press, 2016, www.nap.edu/catalog/23686/a-21st-century-cyber-physical-systems-education.

[64] Mehmet Baygin, Hasan Yetis, Mehmet Karakose, Erhan Akin, An effect analysis of Industry 4.0 to higher education, in: Conference: 2016 15th International Conference on Information Technology Based Higher Education and Training (ITHET), https://doi.org/10.1109/ITHET.2016.7760744.

[65] Mariagiovanna Sami, Miroslaw Malek, Umberto Bondi, Francesco Regazzoni, Embedded systems education: job market expectations, SIGBED Review 14 (1) (October 2016) 22–28.

[66] A. Ziravecka, N. Kunicina, K. Berzina, A. Patlins, Flexible approach to course testing for the improvement of its effectiveness in engineering education, in: Proceedings of the 2015 IEEE 8th International Conference on Intelligent Data Acquisition and Advanced Computing Systems: Technology and Applications (IDAACS), Poland, Warsaw, 24–26 September, 2015, IDAACS'2015 Organizing Committee, Warsaw, 2015, pp. 955–959.

[67] Fadi Kurdahi, Mohammad Abdullah Al Faruque, Daniel Gajski, Ahmed Eltawil, A case study to develop a graduate-level degree program in embedded & cyber-physical systems, SIGBED Review 14 (1) (October 2016) 16–21.

[68] Walid Taha, Yingfu Zeng, Adam Duracz, Xu Fei, Kevin Atkinson, Paul Brauner, Robert Cartwright, Roland Philippsen, Developing a first course on cyber-physical systems, SIGBED Review 14 (1) (October 2016) 44–52.

[69] Albert M.K. Cheng, An undergraduate cyber-physical systems course, in: CyPhy'14, April 14–17, 2014, Berlin, Germany, https://doi.org/10.1145/2593458.2593464.

[70] Daniela Antkowiak, Daniel Luetticke, Tristan Langer, Thomas Thiele, Tobias Meisen, Sabina Jeschke, Cyber-physical production systems: a teaching concept in engineering education, in: 2017 6th IIAI International Congress on Advanced Applied Informatics, https://doi.org/10.1109/IIAI-AAI.2017.35.

[71] N. Kuņicina, A. Žiravecka, J. Čaiko, A. Patļins, L. Ribickis, Research-based approach application for electrical engineering education of bachelor program students in Riga Technical University, in: IEEE Education Engineering Conference (EDUCON 2010), Spain, Madrid, 14–16 April, 2010, IEEE, Piscataway, 2010, pp. 695–700, https://doi.org/10.1109/EDUCON.2010.5492510.

Index

Printed in the United States
By Bookmasters